Soil Contamination
Hazardous Waste and Remediation

World Scientific Series on Advances in Environmental Pollution Management

Print ISSN: 3082-8376
Online ISSN: 3082-8384

Series Editors: Dharmendra Kumar Gupta
(*Ministry of Environment, Forest and Climate Change, India*)
Clemens Walther
(*Leibniz University Hannover, Germany*)

Published

Vol. 4 *Soil Contamination: Hazardous Waste and Remediation*
edited by Dharmendra K Gupta & Clemens Walther

Vol. 3 *Valorization of Resources from Urban Mined Materials*
edited by Dharmendra K Gupta, Pankaj Pathak & Dheeraj Mittal

Vol. 2 *Radionuclide Uptake in Food and Consequences for Humans*
edited by Dharmendra K Gupta & Clemens Walther

Vol. 1 *Lithium and Nickel Contamination in Plants and the Environment*
edited by Masahiro Inouhe, Clemens Walther & Dharmendra K Gupta

World Scientific Series on Advances in
Environmental Pollution Management – Volume 4

Soil Contamination
Hazardous Waste and Remediation

Editors

Dharmendra K Gupta

Clemens Walther

World Scientific

NEW JERSEY • LONDON • SINGAPORE • BEIJING • SHANGHAI • HONG KONG • TAIPEI • CHENNAI • TOKYO

Published by

World Scientific Publishing Co. Pte. Ltd.
5 Toh Tuck Link, Singapore 596224
USA office: 27 Warren Street, Suite 401-402, Hackensack, NJ 07601
UK office: 57 Shelton Street, Covent Garden, London WC2H 9HE

British Library Cataloguing-in-Publication Data
A catalogue record for this book is available from the British Library.

SOIL CONTAMINATION
Hazardous Waste and Remediation

ISBN 978-981-98-2223-2 (hardcover)
ISBN 978-981-98-2224-9 (ebook for institutions)
ISBN 978-981-98-2225-6 (ebook for individuals)

For any available supplementary material, please visit
https://www.worldscientific.com/worldscibooks/10.1142/14557#t=suppl

Desk Editors: Aanand Jayaraman/Amanda Yun

Typeset by Stallion Press
Email: enquiries@stallionpress.com

https://doi.org/10.1142/9789819822249_fmatter

Preface

Polluted soils often represent a hostile environment for living organisms due to the toxicity of pollutants, the loss of soil structure, or the lack of organic matter and nutrients. Soil contamination or soil pollution is generally caused by the presence of man-made chemicals (xenobiotics) or other modifications in the natural soil environment. Hazardous wastes come from a wide variety of sources. Industrial operations, including those in mining, metal processing, and the production of chemicals such as solvents and heavy metals, make up a significant portion. Many agricultural practices include fertilizers, pesticides, and herbicides, many of which contain chemicals that are toxic for a long time. In addition, hazardous substances such as poisons and explosives are produced by military operations and end up in the environment for a long period.

Hazardous waste can travel through a variety of routes, including surface discharge into aquatic systems, volatilization into the atmosphere, and leaching into groundwater. It is characterized by its toxicity, reactivity, and persistence. Its mobility and dispersion are significantly influenced by its chemical makeup, environmental conditions, and disposal methods. If hazardous waste is not appropriately managed throughout storage, transportation, and disposal, there is a substantial risk to both humans and the environment.

There is persistent use of hazardous substances in military equipment and weaponry, as well as in industry and construction materials, which consequently and potentially creates acute and chronic risks for complex biological and ecological systems, leading to a severely contaminated environment of an ecocidal nature. Long-term exposure to contaminated soil can lead to more serious health problems, such as cancer and neurological disorders.

This book, *Soil Contamination from Hazardous Waste: Remediation Measures*, presents the recent research on soil contamination due to hazardous waste, including military operations via metal/heavy metal in the environment, uptake by plants, and finally to the food chain. This book includes a thorough examination of the impacts of hazardous waste on soil contamination and its impact on food and finally on human health. Some chapters are on the physical and chemical impact of hazardous waste on soil and the health risk linked to it. Few chapters are related to military pollution related to metals' transfer factors to plants and food and also its environmental impact.

The editors (Dr. Dharmendra K. Gupta and Prof. Clemens Walther) in person acknowledge all the contributing authors for their precious time, knowledge, and enthusiasm to bring this book into this form.

Dr. Dharmendra K. Gupta
New Delhi, India
Prof. (Dr.) Clemens Walther
Hannover, Germany

https://doi.org/10.1142/9789819822249_fmatter

About the Editors

Dharmendra K. Gupta is Director (S) at the Ministry of Environment, Forest and Climate Change, Government of India. His field of research includes abiotic stress by radionuclides/heavy metals and xenobiotics in plants, antioxidative system in plants, and environmental pollution (radionuclides/heavy metals) remediation through plants (phytoremediation). Dr. Gupta has been awarded several international awards by entities such as the Japan Society for the Promotion of Science (JSPS), the Belgian Science Policy Office (BELSPO), JAE-Doc. (Spain), TWAS-CNPq (Italy), Royal Society India Fellow (U.K.), and MASAV (Israel). Dr. Gupta has published more than 130 international manuscripts and review articles in international and reputed journals, edited 30 books, and served as Book Series editor for 2 book series from Springer, Switzerland, and 1 from Scientific World, Singapore.

Clemens Walther is a Professor at the Leibniz Universität Hannover, Germany, and Head of the Institute of Radioecology and Radiation Protection. He has been Chair of the Nuclear Chemistry Section of the German Chemical Society and is Head of the German Steering Board of the Competence Center Radiation Research (KVSF) since 2017. Since 2015, he has been a member of the German Commission for Radiation Protection and since 2023, Vice President of the German Swiss Society for Radiation Protection. Prof. Walther's past appointments include being Dean of the Faculty of Mathematics and Physics (2019–2021), Head of the European Network on Nuclear and Radiochemistry Education and

Training (2016–2022), Member of the extended governing board of the German Society for Mass Spectrometry (DGMS) (2012–2015), and Head of the mass spectrometry division of the German Physical Society (DPG) (2012–2015). Prof. Walther has published more than 175 international manuscripts and review articles in international and reputed journals and edited 9 books.

https://doi.org/10.1142/9789819822249_fmatter

Contents

https://doi.org/10.1142/9789819822249_0001

Chapter 1

Availability, Distribution, and Mobility of Hazardous Waste Material in the Environment

Anindita Mitra[*,¶], Soumya Chatterjee[†,‖], Anand Dubey[‡], and Dharmendra K Gupta[§]

[*]*Bankura Christian College, Bankura 722101, West Bengal, India*

[†]*Biodegradation Technology Division, Defence Research Laboratory, DRDO, Tezpur 784001, Assam, India*

[‡]*Ind Research & Development House Pvt. Ltd. C-10, Second Floor, Sector-6, Noida-201301, UP, India*

[§]*Ministry of Environment, Forest and Climate Change, Indira Paryavaran Bhavan, Aliganj, Jorbagh Road, New Delhi 110003, India*

[¶]*aninditajashmitra@rediffmail.com*

[‖]*drlsoumya@gmail.com*

Abstract

Hazardous waste refers to waste that if not handled properly poses a danger to human health or environment due to its properties, such as being toxic, ignitable, corrosive, or reactive. It originates from various sources, including industrial processes, commercial activities, hospitals, and household products such as pesticides or cleaning fluids. It requires specific handling, storage, and disposal methods to prevent harm to humans and environment.

Keywords: Distribution, mobility, hazardous waste, environment.

1. Introduction

For almost the last five decades, the issue of hazardous waste management has been a significant concern in a number of nations. If hazardous waste is not appropriately managed throughout storage, transportation, and disposal, there is a substantial risk to both humans and the environment. Industries and industrialization are essential to the economic basis of modern society, and hazardous waste invariably generates hazardous waste. Waste is defined as an item that is thrown away without any expectation of payment for its intrinsic value (Misra and Pandey, 2005). Around 2.1 billion tonnes of municipal solid trash will be produced worldwide in 2023, according to the UNEP's "Global Waste Management Outlook 2024" (https://www.unep.org/resources/global-waste-management-outlook-2024), and waste output is predicted to soar to 3.8 billion tonnes by 2050. India generates about 62 million tons of waste annually, comprising both industrial and municipal waste, with about 170,000 metric tons of municipal solid trash created daily (https://www.trade.gov/market-intelligence/india-solid-waste-management). Hazardous materials are frequently disposed of in the environment, endangering both human health and the ecosystem. These materials are not biodegradable and remain in the environment for a long time. However, to address the growing issue of hazardous compounds in the environment, which is a challenging undertaking because of its many origins, the government and international organizations are taking action. The air, water, and soil are contaminated by hazardous and toxic compounds that come from a variety of sources (Enger and Smith, 2004). The potential health hazards linked to these medications range from minor, temporary discomforts such as headaches and nausea to major issues such as cancer and birth defects, as well as severe mishaps that could result in immediate harm or death, according to Kanagamani *et al.* (2020). Hazardous waste comes from a variety of sources, including discarded consumer items, chemicals, pharmaceuticals, pesticides, personal hygiene products, electronic devices, industrial byproducts, sludge, and waste oils.

By 2050, around 68% of the world's population is expected to reside in urban areas, according to United Nations (UN) projections (Knickmeyer, 2020). A recent research conducted by the United Nations Environment Programme (UNEP) found that the world produces about 400 million tons of hazardous waste annually, or over 60 kg per person annually, and that this amount is steadily rising. In a single generation, the amount of synthetic chemicals produced has grown 400%, from 1 million tons to 400 million tons (The World Counts, 2023). According to Gautam *et al.* (2019), some developed nations generate a lot of hazardous waste, while others produce very little. The United Nations, China, and a few of the EU member nations produce the majority of hazardous waste. There is no commonly accepted system for categorizing hazardous waste. Even within the same country, different regulatory schemes may classify hazardous waste in different ways. The definitions and classifications of hazardous waste and products may vary over time due to climate change (Rabbani *et al.*, 2019). According to its physical, chemical, and infectious properties as well as the potential for injury during handling and disposal, hazardous waste can be divided into different categories (Samant *et al.*, 2018). Workers in laboratories and medical institutions are particularly susceptible to exposure to hazardous materials. It is becoming more widely recognized that a wide range of professions, including those in agriculture, textiles, woodworking, sewage, and composting facilities, have the capacity to generate hazardous waste in different forms. Every day, innumerable hospitals and other medical and scientific institutes across a country generate millions of tons of waste. Furthermore, there is a serious risk to individuals who deal with the handling of these wastes' treatment, disposal, transportation, and storage. In addition, rag pickers throughout gather fluid bottles and other medical waste, including saline bottles, disposable syringes, and intravenous needles, and recycle them back into the market without any kind of cleansing or disinfection. These wastes are hazardous to the environment even at low quantities, and efforts to manage them globally are still in their infancy. In order to handle trash properly and safely, to raise awareness for ongoing vigilance, and to provide information on how to handle hazardous waste found in the environment, preventative measures must be put into place. An in-depth examination of the environmental impacts of hazardous waste produced by industrial, medical, residential, electronic, military, and radioactive sources is the goal of this paper. In order to reduce the harmful consequences of hazardous waste, it also examines existing waste management procedures, regulatory frameworks, and viable remediation techniques.

2. Hazardous Waste Sources

Hazardous wastes come from a wide variety of sources (Fig. 1). Industrial operations, including those in mining, metal processing, and the production of chemicals such as solvents and heavy metals, make up a significant portion. Many agricultural practices include fertilizers, pesticides, and herbicides, many of which contain chemicals that are toxic for a long time. A significant amount originates from domestic waste, which includes items that frequently include hazardous chemicals, such as paints, batteries, electronics, and cleaning supplies (Tejaswini *et al.*, 2023). If not adequately managed, pharmaceutical and biomedical waste produced by healthcare facilities can present major hazards to both human health and the environment. Furthermore, hazardous substances such as poisons and explosives are produced by military operations and end up in the environment for a long period. The origins and characteristics of hazardous waste production highlight the necessity of all-encompassing waste management procedures and regulations to mitigate their effects.

2.1. *Industrial waste*

Industrialization significantly boosts a nation's economy and, consequently, its residents' level of living. There are major repercussions as a result of the excessive generation of hazardous waste and inadequate management. Industrial waste can be classified as solid, liquid, or gaseous. The Central Pollution Control Board (CPCB) of India classified 64 industries—including small, medium, and big enterprises—as "Red

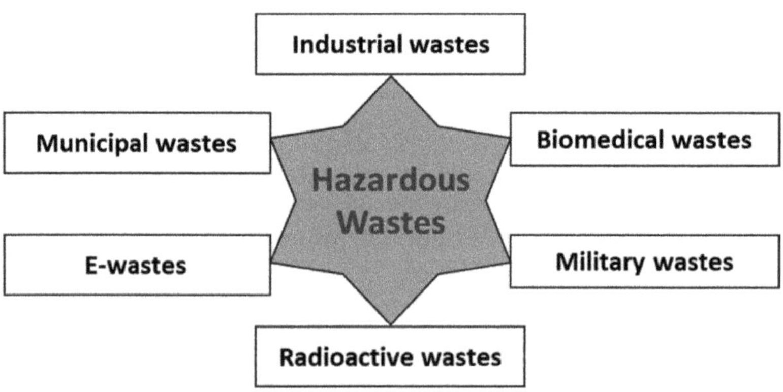

Fig. 1. Types of hazardous wastes.

Category" due to their emissions of hazardous waste. Waste characterization is the first step in any successful industrial waste management plan (Sharma and Lewis, 1994). Prior to developing measures for waste control and treatment, it is essential to ascertain whether the trash contains hazardous components (Hoornweg, 2000). One of the simplest ways to evaluate industrial waste is to look at industries that are known to have a significant ability to produce hazardous wastes, such as lead acid batteries, tanning, petroleum, electroplating, and pesticides.

Once hazardous waste has been identified, it can be categorized based on the degree of risk it poses. Three different categories can be used to group these wastes (Misra and Pandey, 2005).

High-risk industrial waste: These substances contain significant amounts of substances with high potential for toxicity, mobility, persistence, and/or bioaccumulation. Heavy metals, insecticides, hydrocarbons (vinyl chloride and benzene), and organic compounds (PCBs and dioxins) are a few examples. Intermediate-risk wastes include metal hydroxide sludges, except for Cr^{6+}, which is categorized as a high-risk waste because of its high toxicity. The mobility and solubility of the hazardous metals in this category are comparatively modest. Low rank wastes are mostly generated in high volume. This group includes the industrial waste from pulp and paper, sugar mills, textiles, tanneries, food processing, oil palm, distilleries, and dairy farms.

2.2. *Medical waste*

Biomedical waste is any waste generated during the diagnosis, treatment, or immunization of a human or animal. As stated by Gutberlet and Uddin (2017), "Waste generated while diagnosing, treating, or immunizing people or animals". The uncontrolled handling of biomedical waste poses a serious threat to human health and safety, as well as the environment for present and future generations (Chartier *et al.*, 2014). Biomedical wastes include things such as sharps, radioactive waste, chemical waste, pharmaceutical waste, pathological waste, infectious waste, pressurized containers, and heavy metal contents (Sahiledengle, 2019; Das *et al.*, 2021). Biomedical concerns can arise from a variety of exposure routes, including cuts or pricks, direct contact with skin or mucosal membranes, inhalation, or ingestion of infectious material (He *et al.*, 2022; Sefouhi *et al.*, 2013). In recent years, the expansion of medical and surgical procedures, the use of plastic disposables, and the growth of healthcare facilities have all contributed to an unprecedented amount of biomedical waste. According to the World Health Organization (WHO) research in 22 impoverished nations, between 18% and 64% of health care facilities (HCF) do not use suitable waste disposal practices (WHO, 2011). The annual production of biomedical waste in India ranges from 0.5 to 2.0 kg/bed/day, or around 0.33 million tons (Mathur *et al.*, 2011). The increasing volume of biomedical waste brought on by improvements in healthcare calls for the adoption of sustainable waste management techniques, which include appropriate methods for disposal, treatment, and segregation. The risks associated with biomedical waste can be significantly decreased by raising awareness, enhancing infrastructure, and enforcing strict adherence to waste management laws, fostering a safer and healthier future for communities everywhere.

2.3. *Waste from the home*

Over two-thirds of the municipal solid waste (MSW) stream is made up of household waste that comes from domestic sources, and a sizable amount of it is dumped in landfills worldwide (Eurostat, 2011; Slack *et al.*, 2005). Household hazardous waste (HHW) is a small portion, typically 1% by weight. "Leaf over home items that contain corrosive, poisonous, ignitable, or reactive components" is how US law defines HHW (USEPA, 2005). "Wastes that could potentially enhance the hazardous characteristics of municipal solid waste when land filled, burned, or composted" is how the term "HHW" is defined in Europe (Gendebien *et al.*, 2002). The following traits are present in household hazardous materials: toxicity, corrosiveness, flammability, and reactivity (Wessel, 2004). Solvents, antifreeze, pesticides, oil-based paints, batteries, and leftover pharmaceuticals are just a few examples of common household goods that display these characteristics (Bound *et al.*, 2006; EPA, 2008; Weidenhame *et al.*, 2010).

2.4. *Electronic waste*

The need for electronic devices, such as televisions, cell phones, computers, hardware, air conditioners, refrigerators, and other electronic equipment for home, office, and personal use, has grown dramatically due to technological advancements, quick industrial development, economic expansion, and changing lifestyles (Needhidasan *et al.*, 2014; Iyer *et al.*, 2025). The high demand for electronic products has resulted in a significant amount of electronic garbage, or "e-waste," on a global scale (Roy *et al.*, 2022). Both hazardous and non-hazardous elements are present in e-waste, including valuable components such as steel and iron (50%) and other metals (13%), plastics (21%), and other materials, including glass, ceramics, wood, and rubber (16%) (Wath *et al.*, 2011). There is a growing use of platinum (Pt), gold (Au), cadmium (Cd), silver (Ag), palladium (Pd), zinc (Zn), lead (Pb), copper (Cu), cobalt (Co), nickel (Ni), and rare earth minerals such as yttrium (Y), lanthanum (La), cerium (Ce), and neodymium (Nd) in electronic devices such as printed circuit boards (PCBs), computers, smart phones, televisions, printers, refrigerators, telecommunication servers, washing machines, photocopiers, coffee makers, and photocopiers (Graedel and Lifset, 2016; Bhanu *et al.*, 2025). Globally, more than 62 million tons of electronic trash were created in 2022 (Gupta *et al.*, 2025). According to Balde *et al.* (2024), just 22.3% of the waste was legally collected and recycled. In addition to being stored in homes and warehouses, millions of tons of e-waste are improperly handled, thrown out, and illegally exported. Up to 1000 different chemicals, including known neurotoxins such as lead, can be released into the environment when e-waste is not recycled properly (Widmer *et al.*, 2005). Due to their developmental circumstances and exposure routes, toddlers and pregnant women are more vulnerable. According to the International Labour Organization (ILO), 16.5 million children worked in the industrial sector, which includes the subsector of waste processing, in 2020 (Child labor: global estimates, 2020). As a result, poor e-waste management causes these dangerous materials to leak and leak into the soil, water, and air, which kills some aquatic plants and animals (Ankit *et al.*, 2021). These compounds may have negative effects on the human heart, kidneys, brain, liver, reproductive system, and skeletal system if absorbed by plants, drinking water, or direct inhalation (Kumar *et al.*, 2017).

2.5. *Waste from the military*

Materials that have a high potential for detonation are considered explosive. Three types of potential energy can be found in explosives: nuclear, compressed gas, and chemical energy. Both military and civilian applications require these potent compounds. Even though the word "explosive" conjures up images of pain and violence, its constituents have numerous significant applications in metallurgy (metal bonding), engineering, mining, quarrying, construction, aerospace (rocket propulsion), and other fields (Keegan, 1994; Chatterjee *et al.*, 2017). Military forces most frequently use octahydro-1,3,5,7-tetranitro-1,3,5,7-tetrazocine (HMX), hexahydro-1,3,5-trinitro-1,3,5-triazine (RDX), and 2,4,6-trinitrotoluene (TNT). Perchlorate formulations such as nitroglycerin (NG), nitrocellulose (NC), nitroguanidine (NQ), 2,4-dinitrotoluene (DNT), and others are frequently used in rocket, missile, and firearm propellants (Pichtel, 2012). According to Pennington and Brannon (2002), Pichtel (2012), and Charles *et al.* (2014), there is a substantial negative environmental impact from the chemicals and toxic substances related to unexploded ordnance (UXO) and post-explosion by-products, especially from industrial operations, military drills, and war. Due to possible hazards to human health and the environment, the U.S. Environmental Protection Agency (USEPA) has designated a number of locations with elevated levels of contamination in the air, water, and soil (Pennington and Brannon, 2002; Juhasz and Naidu, 2007; USEPA, 2014; Pichtel, 2012; Panz *et al.*, 2013; Charles *et al.*, 2014; Noctor, 2015). The majority of munitions are inherently hazardous and have the potential to cause cancer (Via and Zinnert, 2016; Eum *et al.*, 2016).

However, little research has been done on these traits and their rehabilitation potential (Charles *et al.*, 2014; Via and Zinnert, 2016). For example, long-term exposure to 2,4,6-trinitrotoluene (TNT), a frequently used component of weapons, may have serious effects on the environment, especially on marine and coastal ecosystems (Nipper *et al.*, 2004). In addition to their possible mutagenic and/or carcinogenic qualities, some studies show that substances such as TNT and its byproducts can also negatively impact the liver and spleen (Nipper *et al.*, 2005; Pichtel, 2012; Charles *et al.*, 2014). Serious symptoms such as convulsions, vertigo, vomiting, and unconsciousness have been reported in industry workers in the US and Europe after being exposed to RDX (Best *et al.*, 1999; Pichtel, 2012). Similarly, NG products can cause circulatory collapse, cyanosis, headaches, convulsions, nausea, and even death (Sittig, 1991). For TNT, RDX, and HMX, the USEPA has set lifetime exposure limits in drinking water of 2.0, 2.0, and 400 mg L_{-1}, respectively (Crockett *et al.*, 1999; Pichtel, 2012). Effective remediation techniques must be used, even if human society frequently depends on the useful applications of explosives. These procedures are crucial for safeguarding ecosystems against the movement and substances released during explosions, as well as the long-term contamination brought on by the inappropriate handling and disposal of unexploded ordnance (Darrach *et al.*, 1998; Pichtel, 2012; Charles *et al.*, 2014).

Incineration, advanced oxidation processes (AOP), and alkaline hydrolysis are examples of standard strategies for dealing with HMX pollution in soil and water (Nagar *et al.*, 2021). Two techniques that have been researched for the removal of HMX from contaminated environments include phytoremediation and microbial degradation. Numerous studies suggest that some aquatic and terrestrial plants may absorb HMX; however, the breakdown process is sluggish and usually requires acclimatization (Nagar *et al.*, 2021). Catharanthus roseus and Myriophyllum aquaticum showed little HMX transformation, and prolonged exposure allowed for more removal (Medina *et al.*, 1998). Although microbial degradation offers a potential treatment, its efficiency is hampered by HMX's chemical stability, which causes a sluggish degradation process. Within biogeochemical cycles, certain bacteria can break down HMX, yet enzymes that permit full breakdown are rarely present. According to Zhang *et al.* (2013), microbial granules in an aerobic reactor successfully decomposed HMX in wastewater, eliminating 97.57% of the organic matter.

2.6. *Radioactive waste*

Materials that display radioactivity are included in the category of hazardous rubbish known as "radioactive waste." This includes any substance that is judged to have no further use and is either naturally radioactive or has been impacted by radiation. Nuclear power, nuclear medicine, nuclear research, coal and rare-earth mining, manufacturing, construction, and nuclear weapons reprocessing are among the businesses that produce radioactive waste (Zhang *et al.*, 2019). Since radioactive waste poses a substantial risk to human and environmental health, it must be managed effectively to ensure that radiation exposure stays within defined limits and guidelines (Poškas *et al.*, 2019; Walther and Gupta, 2015).

Nuclear energy has been widely employed to create electricity in recent decades. Given the existing circumstances, there will not be a major decline in uranium consumption in the years to come, implying that the creation of radioactive waste is on the rise (Ma *et al.*, 2019). Furthermore, eating tainted food cultivated in places that are irrigated from radionuclide-contaminated water might cause radionuclides to infiltrate the food chain (Walther and Gupta, 2015; Gupta and Walther, 2025). Solutions that are acceptable, efficient, and ecologically friendly are needed to address the issue of radioactive contamination in natural water. By employing natural materials to support rehabilitation and lessen radionuclide migration, a green technological paradigm can be put into practice (Gupta *et al.*, 2018). Many materials, including cellulose, lignin, humic acids, activated carbons, and phytoplankton, are employed as green sorbents to remove radioactive contaminants. For a number of radionuclides, including radon, uranium, thorium, plutonium, strontium, cesium, and lanthanides, activated carbons have been demonstrated to be effective sorbents. Since the radionuclide can move far from the source of emission once polluted, the radioactive contaminated zone can expand (Gupta and Walther, 2014). Consequently, managing and preserving radioactive waste and wastewater has become extremely difficult. Numerous techniques, including subterranean burial, extraterrestrial disposal, and ocean disposal, have been used to get rid of high-level radioactive waste (Brunnengräber, 2019).

3. Hazardous waste mobility

Hazardous waste mobility is a significant environmental and public health concern because it determines the extent to which hazardous wastes move through various ecosystems and have the potential to contaminate soil, water, and air. Hazardous waste can travel through a variety of routes, including surface discharge into aquatic systems, volatilization into the atmosphere, and leaching into groundwater. It is characterized by its toxicity, reactivity, and persistence. Its mobility and dispersion are significantly influenced by its chemical makeup, environmental conditions, and disposal methods. Environmental degradation can also be accelerated by surface runoff and erosion, which can encourage the movement of hazardous waste from polluted areas to nearby aquatic bodies. Temperature, microbial activity, pH levels, and soil characteristics all affect the degree of mobility and can either help or hinder the breakdown and transport of contaminants. Developing effective waste management strategies, legal frameworks, and remediation techniques to reduce environmental and health risks requires an understanding of these complex interactions.

Garbage has been dumped in urban areas since the beginning of human civilization, which has led to urban pollution and the rise of deadly diseases (Martin, 2008). Municipal and town administrations have started collecting rubbish and disposing of it in landfills or aquatic areas in an effort to improve the quality of life. The main problem associated with landfilling garbage is the production of LFL, a toxic wastewater type that can contaminate groundwater. The scientific and technological community began to recognize the risks and detrimental effects of landfilling on the environment in the twenty-first century (Quan *et al.*, 2020; Sun *et al.*, 2022). Environmental degradation is a result of the significant shift in waste composition from biodegradable materials to hazardous waste and non-biodegradable polymers (Ward, 2011; Kumar *et al.*, 2023). According to an analysis of environmental research conducted in Brazil, landfills and dumpsites have the potential to contaminate soil, surface water, and groundwater with heavy metals, lead, and coliforms. Furthermore, it has been shown that India's informal recycling of waste—especially electronic waste—causes environmental contamination and poses health risks to both residents and workers due to exposure to dangerous chemicals, including heavy metals and polychlorinated biphenyls. It has been demonstrated that children exposed to landfills suffer from oxidative stress-mediated damage to macromolecules, which can have serious health repercussions. Many developed nations have enacted legislation governing landfills in an effort to stop and lessen environmental harm (Abunama *et al.*, 2021). Reduce, reuse, and recycle waste whenever possible to keep it out of landfills (Can-Güven *et al.*, 2021; Ishii *et al.*, 2022). Furthermore, MSW LFL is a highly concentrated organic effluent with a complicated composition. This contributes significantly to pollution and endangers the quality of surface and groundwater. Before being released into the environment, leachates must undergo extensive treatment (Chelliapan *et al.*, 2020). Methane (CH_4) and hydrogen sulfide are two off-gases released by landfills that negatively impact air quality. Reducing greenhouse gas emissions through better waste management techniques can help mitigate health and environmental issues as well as landscape degradation (Giersc *et al.*, 2018).

Human health and the environment are seriously harmed by clinical waste (CW). A suitable waste management system is necessary for the efficient disposal of hazardous chemical waste. Incineration is the best way to reduce the volume of such trash (Abd El-Salam, 2010; Retnaraj *et al.*, 2021). Since metals are left over after incineration, using a landfill to dispose of clinical waste ash could contaminate groundwater (Kaur *et al.*, 2019; Ababneh *et al.*, 2020). Moving CW ashes to a landfill is the most common method of disposal. Recycling technologies and the repurposing of ash in various systems have emerged as a result of spatial constraints and the high costs of land disposal (Ghazali *et al.*, 2022). A solid waste byproduct of burning clinical waste, clinical waste incinerator fly ash (CWIFA), typically includes leachable heavy metals and presents significant environmental concerns when repurposed. The U.S. Environmental Protection Agency's (USEPA) metal threshold must be below the standard criterion for CWIFA removal in landfills (Ababneh *et al.*, 2020). Therefore, effective mitigation strategies are crucial for lowering soil, water, and air pollution. These strategies include advanced treatment technology, landfill restrictions, and sustainable waste disposal practices. Additionally, promoting recycling, resource recovery, and trash reduction can help lessen the growing amount of hazardous and municipal solid waste. In order to preserve ecosystems and advance a healthy future, an integrated approach that blends scientific advancement with governmental action is essential.

4. Hazardous Waste Distribution

A major environmental and public health concern is the distribution of hazardous waste, which encompasses the manufacturing, transportation, treatment, storage, and disposal of items that seriously endanger human populations and ecosystems. 11.2 billion tons of solid garbage are collected each year worldwide, according to the United Nations Environment Programme (UNEP), and the waste market sector as a whole, including collection and recycling, is valued at US$ 410 billion (UNEP, 2012, 2020). This estimate excludes the informal sector, which in low-income countries recycles 80% of waste (Rucevska *et al.*, 2015). This garbage can contaminate the air, water, and soil if it is not adequately managed, which will increase environmental deterioration and lead to long-term health issues. Understanding the temporal and spatial distribution of hazardous waste is essential for creating effective waste management programs, lowering hazards, and ensuring regulatory compliance. In recent years, environmental problems have been made worse by population increase, industrial and technological advancements, and governments' concerns about hazardous waste management (HWM) systems have grown.

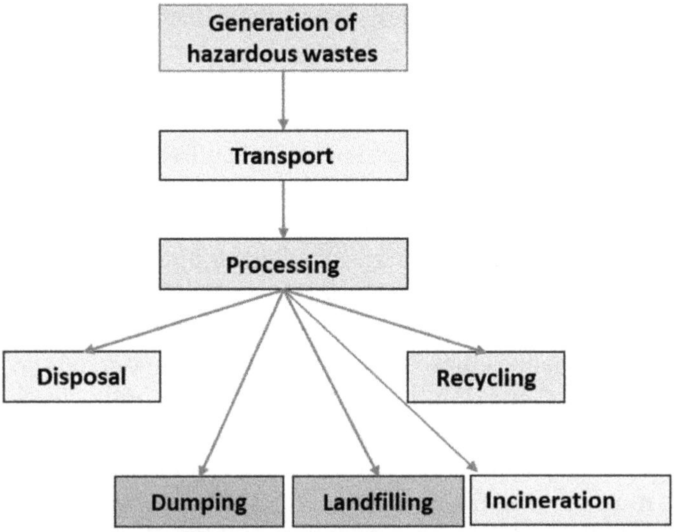

Fig. 2. Scheme for hazardous waste management.

Hazardous waste management encompasses a wide range of activities, including the collection, transportation, recycling, treatment, and disposal of hazardous waste (Fig. 2), as well as the identification of suitable routes for the collection and delivery of hazardous material and waste residues between facilities (Zabihian-Bisheh *et al.*, 2024). Since there are numerous varieties of physically and chemically hazardous waste, choosing suitable sites for treatment, recycling, and disposal facilities, as well as allocating hazardous waste to these facilities, is a crucial and sometimes challenging task in the hazardous waste management (HWM) system (Nema and Gupta, 1999). Since it must simultaneously assess multiple points of view, such as social, economic, and environmental issues, developing an effective HWM system is more difficult (Jacobs and Warmerdam, 1994). To ensure human safety, a system for the sustainable management of hazardous waste (SHWM) needs to be in place. A waste management system's (WMS) ability to manage sustainability determines its effectiveness (Asase *et al.*, 2009). However, it must be acknowledged that sustainability necessitates preserving the environment, achieving economic efficiency, and gaining the support of society. One of the main ways these facilities attempt to lessen the environmental impact of solid waste is by reducing open dumping. Although these facilities were operating, they encountered issues that ultimately led to their demise. Their failure and eventual demise were largely caused by their inefficiency and excessive operating costs, which were brought on by the incorrect implementation of sustainability. As a result, SHWMs need to be affordable, ecologically sustainable, and socially acceptable (McDougall *et al.*, 2008). Facilities for incineration and composting in Turkey (Kanat, 2010) and Delhi (Talyan *et al.*, 2008) serve as illustrative examples.

Population growth in urban areas is becoming a concern worldwide. Delhi, with a notable population growth rate of 3.85% between 1991 and 2001—nearly twice the national average, is one of India's most urbanized and densely populated cities. Delhi's position as a commercial hub has led to employment creation and urban expansion, both of which have increased the production of municipal solid waste (MSW) (Zabihian-Bisheh *et al.*, 2024). According to a report by Talyan *et al.* (2008), the city generates roughly 13,000 tons of municipal solid garbage every day and approximately 90% of rubbish gets collected (https://www.thehindu.com/sci-tech/energy-and-environment/delhi-mounting-waste-crisis-solid-waste-management/article68179968.ece). Nevertheless, municipal solid waste management is still a neglected area in many a country including India, as most of the collected wastes ending up in open landfills or on the streets. Composting, the only treatment option used, accounts for only 9% of the collected waste, with the balance being dumped in uncontrolled locations. These landfills' lack of leachate and gas collection systems contributes significantly to groundwater contamination and air pollution, which includes greenhouse gas emissions. Due to a number of operational challenges, composting plants operate at a reduced capacity, increasing the need for landfills. Three competing goals must be balanced in order to achieve sustainability: social progress, environmental preservation, and economic profit. Environmental preservation is emphasized as one of these objectives in this definition, which was proposed by the World Summit on Sustainable Development (WSSD) (White and Lee, 2009).

5. Final Thoughts

The management of hazardous waste is becoming a more challenging issue that calls for immediate, innovative, and sustainable solutions. With the growth of industry, urbanization, and technological breakthroughs, hazardous waste becomes more complicated and abundant, posing major risks to public health and the environment. The complex characteristics of hazardous waste, including industrial wastes, biomedical waste, and radioactive and electronic contamination, need for an all-encompassing, science-based strategy that combines strict rules, cutting-edge treatment technology, and international cooperation.

Despite advancements in waste management laws, poor infrastructure and slack enforcement remain significant challenges. Resource recovery, safe disposal, and effective segregation are urgent needs that necessitate proactive steps, such as increased public awareness, frameworks for the circular economy, and advancements in green chemistry. Moreover, interdisciplinary research and policy frameworks need to give priority to environmentally friendly waste remediation techniques that employ contemporary recycling technology, microbial degradation, and phytoremediation.

A sustainable future requires considerable collaboration among various stakeholders. Governments, corporations, scientists, and communities must collaborate in order to prevent hazardous waste problems and transition to ecologically friendly waste management techniques. The ecosystem and public health may suffer long-term consequences if hazardous waste is allowed to grow out of control. The time has come to act to make the planet a more resilient, cleaner, and safer place.

Acknowledgments

SC conveying sincere thanks to the Director, DRL, for the preparation of the MS.

Bibliography

Ababneh, A., Al-Rousan, R., Gharaibeh, W., and Abu-Dalo, M. (2020). Recycling of pre-treated medical waste fly ash in mortar mixtures. *Journal of Material Cycles and Waste Management*, 22(1), 207–220.

Abd El-Salam, M.M. (2010). Hospital waste management in El-Beheira governorate, Egypt. *Journal of Environmental Management*, 91(3), 618–629.

Abunama, T., Moodley, T., Abualqumboz, M., Kumari, S., and Bux, F. (2021). Variability of leachate quality and polluting potentials in light of leachate pollution index (LPI) – A global perspective. *Chemosphere*, 282, 131119.

Ankit, Saha, L., Kumar, V., Tiwari, J., Sweta, Rawat, S., Singh, J., and Bauddh, K. (2021). Electronic waste and their leachates impact on human health and environment: global ecological threat and management. *Environmental Technology & Innovation*, 24, 102049.

Asase, M., Yanful, E.K., Mensah, M., Stanford, J., and Amponsah, S. (2009). Comparison of municipal solid waste management systems in Canada and Ghana: A case study of the cities of London, Ontario, and Kumasi, Ghana. *Waste Management*, 29(10), 2779–2786.

Baldé, C.P., Kuehr, R., Yamamoto, T., McDonald, R., D'Angelo, E., Althaf, S., *et al.* (2024). *The Global E-waste Monitor 2024*. International Telecommunication Union, United Nations Institute for Training and Resources, Bonn, Geneva. https://ewastemonitor.info/.

Best, E.P.H., Sprecher, S.L., Larson, S.L., *et al.* (1999). Environmental behaviour of explosives in groundwater from the Milan army ammunition plant in aquatic and wetland plant treatments. Removal, mass balances and fate in groundwater of TNT and RDX. *Chemosphere*, 38, 3383–3396.

Bhanu, M.L.S., Mitra, A., Gupta, D.K., and Chatterjee, S. (2025). Recovery of elements using green plants for potential valorization of waste metals. In: Gupta, D.K., Mittal, D., Pathak, P. (Eds.), *Valorization of Resources from Urban Mined Materials*. World Scientific Co. Pte. Ltd., Singapore, pp. 173–220.

Bound, J.P., Kitsou, K., and Voulvoulis, N. (2006). Household disposal of pharmaceuticals and perception of risk to the environment. *Environmental Toxicology and Pharmacology*, 21, 301–307.

Brunnengräber, A. (2019). The wicked problem of long term radioactive waste governance. In: *Conflicts, Participation and Acceptability in Nuclear Waste Governance*. Springer VS, Wiesbaden, pp. 335–355.

Can-Güven, E., Guvenc, S.Y., and Varank, G. (2021). Sequential coagulation and heat activated persulfate-peroxide binary oxidation process for landfill leachate treatment. *Journal of Water Process Engineering*, 42, 102202.

Charles, P.T., Adams, A.A., Deschamps, J.R., *et al.* (2014). Detection of explosives in a dynamic marine environment using a moored TNT immunosensor. *Sensors*, 14, 4074–4085.

Chartier, Y., Emmanuel, J., Pieper, U., Prüss, A., Rushbrook, P., Stringer, R., *et al.* (2014). *Safe Management of Wastes from Health-Care Activities* (2nd ed.). WHO Blue Book, Geneva, Switzerland.

Chatterjee, S., Deb, U., Datta, S., Walther, C., and Gupta, D.K. (2017). Common explosives (TNT, RDX, HMX) and their fate in the environment: Emphasizing bioremediation. *Chemosphere*, 184, 438–451.

Chelliapan, S., Arumugam, N., Din, M.F.M., Kamyab, H., and Ebrahimi, S.S. (2020). Anaerobic treatment of municipal solid waste landfill leachate. In: Singh, L., Yousuf, A., Mahapatra, D.M. (Eds.), *Bioreactors: Sustainable Design and Industrial Applications in Mitigation of GHG Emissions*. Elsevier, pp. 175–193.

Child Labour: Global Estimates 2020, Trends and the Road Forward. (2021). International Labour Organization, Geneva. https://www.ilo.org/ipec/Child-labourstatisticsSIMPOC/lang--en/index.htm.

Crockett, A., Craig, H., and Jenkins, T. (1999). *Field Sampling and Selecting On-site Analytical Methods for Explosives in Water.* Report EPA/600/S-99/002. US Environmental Protection Agency, Washington, DC.

Darrach, M.R., Chutjian, A., and Plett, G.A. (1998). Trace explosives signatures from World War II unexploded undersea ordnance. *Environmental Science & Technology*, 32, 1354–1358.

Das, A.K., Islam, M.N., Billah, M.M., and Sarker, A. (2021). COVID-19 pandemic and healthcare solid waste management strategy – A mini-review. *Science of the Total Environment*, 778, 146220.

Enger, E.D. and Smith, B.F. (2004). *Environmental Science: A Study of Interrelationships.* Edward E. Bartell, California, USA.

EPA. (2008). *National Hazardous Waste Management Plan 2008–2012.* Environmental Protection Agency, Ireland.

Eum, J., Kwak, J., Kim, H.J., *et al.* (2016). 3D Visualization of developmental toxicity of 2,4,6-trinitrotoluene in Zebrafish embryogenesis using light-sheet microscopy. *International Journal of Molecular Sciences*, 17(11), E1925.

Eurostat. (2011). Environment in the EU27, recycling accounted for a quarter of total municipal waste treated in 2009. *Eurostat News Release*, 37/2011-8 March 2011.

Gautam, P., Kumar, S., and Lokhandwala, S. (2019). Advanced oxidation processes for treatment of leachate from hazardous waste landfill: A critical review. *Journal of Cleaner Production*, 237, 117639. https://doi.org/10.1016/J.JCLEPRO.2019.117639.

Gendebien, A., Leavens, A., Blackmore, K., Godley, A., Lewin, K., Franke, B., and Franke, A. (2002). *Study on Hazardous Household Waste (HHW) with a Main Emphasis on Hazardous Household Chemicals (HHC).* Final Report, European Commission, Directorate General Environment.

Ghazali, E., Johari, M.A., Fauzi, M.A., and Nor, N.M. (2022). An overview of characterisation, utilisation, and leachate analysis of clinical waste incineration ash. *International Journal of Environmental Research*, 16(5), 69.

Giersc, M., Montevecchi, F., and Neubauer, C. (2018). *Waste Management in Europe: Main Problems Identified in EU Petitions and Best Practices.* http://publications.europa.eu/resource/cellar/8af55d22-3946-11e8-b5fe01aa75ed71a1.0001.03/DOC_1. Accessed 13 July 2022.

Graedel, T.E. and Lifset, R.J. (2016). Industrial ecology's first decade. In: Clift, R., Druckman, A. (Eds.), *Taking Stock of Industrial Ecology.* Springer International Publishing, Cham.

Gupta, D.K. and Walther, C. (2014). *Radionuclide Contamination and Remediation Through Plants.* Springer, Heidelberg.

Gupta, D.K., Voronina, A.V., Semenishchev, V.S., and Chatterjee, S. (2018). Green adsorbents for radioactive pollutants removal from natural water. In: Crini, G., Lichtfouse, E. (Eds.), *Green Adsorbents for Pollutant Removal: Innovative Materials*. Springer Publication, Germany, pp. 377–396.

Gupta, D.K., Mittal, D., and Pathak, P. (2025). *Valorization of Resources from Urban Mined Materials*. World Scientific Co. Pte. Ltd., Singapore.

Gupta, D.K. and Walther, C. (2025). *Radionuclide Uptake in Food and Consequences for Humans*. World Scientific Co. Pte. Ltd., Singapore.

Gutberlet, J. and Uddin, S.M.N. (2017). Household waste and health risks affecting waste pickers and the environment in low-and middle-income countries. *International Journal of Occupational and Environmental Health*, 23(4), 299–310.

He, L., Michailidou, F., Gahlon, H.L., and Zeng, W. (2022). Hair dye ingredients and potential health risks from exposure to hair dyeing. *Chemical Research in Toxicology*, 35(6), 901–915.

Hoornweg, D. (2000). What a waste – Solid waste management in Asia. *UNEP Industry and Environment*, 65–70.

Ishii, K., Sato, M., and Ochiai, S. (2022). Prediction of leachate quantity and quality from a landfill site by the long short-term memory model. *Journal of Environmental Management*, 310, 114733.

Iyer, A., Nambiar, K.S., Gupta, D.K., and Murugan, S. (2025). Metals recovery from electrical and electronic waste. In: Gupta, D.K., Mittal, D., Pathak, P. (Eds.), *Valorization of Resources from Urban Mined Materials*. World Scientific Co. Pte. Ltd., Singapore, pp. 15–34.

Jacobs, T.L. and Warmerdam, J.M. (1994). Simultaneous routing and siting for hazardous-waste operations. *Journal of Urban Planning and Development*, 120(3), 115–131.

Juhasz, A.L. and Naidu, R. (2007). Explosives: fate, dynamics, and ecological impact in terrestrial and marine environments. *Reviews of Environmental Contamination and Toxicology*, 191, 163–215.

Kanagamani, K., Geethamani, P., and Narmatha, M. (2020). Hazardous waste management. In: *Environmental Issues and Sustainable Development*. IntechOpen.

Kanat, G. (2010). Municipal solid-waste management in Istanbul. *Waste Management*, 30(8-9), 1737–1745.

Kaur, H., Siddique, R., and Rajor, A. (2019). Influence of incinerated biomedical waste ash on the properties of concrete. *Construction and Building Materials*, 226, 428–441.

Keegan, J. (1994). *A History of Warfare*. Alfred A. Knopf, New York.

Knickmeyer, D. (2020). Social factors influencing household waste separation: A literature review on good practices to improve the recycling performance of urban areas. *Journal of Cleaner Production*, 245, 118605.

Kumar, A., Holuszko, M., and Espinosa, D.C.R. (2017). E-waste: An overview on generation, collection, legislation and recycling practices. *Resources, Conservation and Recycling*, 122, 32–42.

Kumar, S.S., Kumar, A., Malyan, S.K., Ghosh, P., Kumar, M., Kapoor, R., Agrawal, A.K., Kumar, S., Kumar, V., and Singh, L. (2023). Landfill leachate valorization: A potential alternative to burden off resources and support energy systems. *Fuel*, 331, 125911.

Ma, Z., Gamage, R.P., Rathnaweera, T., and Kong, L. (2019). Review of application of molecular dynamic simulations in geological high-level radioactive waste disposal. *Applied Clay Science*, 168, 436–449.

Mandal, A., Thakur, J.K., Sarkar, A., Saha, M., Solanki, M.K., Rudrashetti, A.P., Singh, A.B., and Patra, A.K. (2023). Efficacy of microbial endophytes in bioremediation: Current research and future outlook. In: *Microbial Endophytes and Plant Growth*, 1, 283–295.

Martin, A.M. (2008). Sewage and the city: Filth, smell, and representations of urban life in Moscow, 1770–1880. *Russian Review*, 67, 243–274.

Mathur, V., Dwivedi, S., Hassan, M., and Misra, R. (2011). Knowledge, attitude, and practices about biomedical waste management among healthcare personnel: A cross-sectional study. *Indian Journal of Community Medicine*, 36, 143–145.

McCarthy, S., Moriarty, J., O'Riordan, D., and O'Leary, G. (2010). *Focus on Landfilling in Ireland.* Environmental Protection Agency, Ireland. https://epa.ie/publications/compliance--enforcement/waste/EPA_Focus_on_Landfilling_Ireland.pdf. Accessed 12 July 2022.

McDougall, F.R., White, P.R., Franke, M., and Hindle, P. (2008). *Integrated Solid Waste Management: A Life Cycle Inventory*. John Wiley & Sons.

Medina, V.F., Revera, R., Larson, S.L., and McCutcheon, S.C. (1998). Phytoreactors show promise in treating munitions contaminants. *Soil & Groundwater Clean Up*, 19–24.

Misra, V. and Pandey, S.D. (2005). Hazardous waste, impact on health and environment for development of better waste management strategies in future in India. *Environment International*, 31(3), 417–435.

Nagar, S., Anand, S., Chatterjee, S., Rawat, C.D., Lamba, J., and Rai, P.K. (2021). A review of toxicity and biodegradation of octahydro-1, 3, 5, 7-tetranitro-1, 3, 5, 7-tetrazocine (HMX) in the environment. *Environmental Technology & Innovation*, 23, 101750.

Needhidasan, S., Samuel, M., and Chidambaram, R. (2014). Electronic waste - an emerging threat to the environment of urban India. *Journal of Environmental Health Science and Engineering*, 12, 36.

Nema, A.K. and Gupta, S. (1999). Optimization of regional hazardous waste management systems: an improved formulation. *Waste Management*, 19(7–8), 441–451.

Nipper, M., Carr, R.S., Lotufo, G.R., *et al.* (2005). Fate and effects of picric acid and 2,6-DNT in marine environment: Toxicity of degradation products. *Marine Pollution Bulletin*, 50, 1205–1217.

Noctor, G. (2015). Lighting the fuse on toxic TNT. *Science*, 349, 1052–1053.

Panz, K., Miksch, K., and Sojka, T. (2013). Synergetic toxic effect of an explosive material mixture in soil. *Bulletin of Environmental Contamination and Toxicology*, 91, 555–559.

Pennington, J.C. and Brannon, J.M. (2002). Environmental fate of explosives. *Thermochimica Acta*, 384, 163–172.

Pichtel, J. (2012). Distribution and fate of military explosives and propellants in soil: a review. *Applied and Environmental Soil Science*, Article ID 617236. http://dx.doi.org/10.1155/2012/617236.

Poškas, P., Kilda, R., Šimonis, A., Jouhara, H., and Poškas, R. (2019). Disposal of very low-level radioactive waste: Lithuanian case on the approach and long-term safety aspects. *Science of the Total Environment*, 667, 464–474.

Quan, X., Hu, R., Chang, H., Tang, X., Huang, X., Cheng, C., Zhong, N., and Yang, L. (2020). Enhancing microalgae growth and landfill leachate treatment through ozonization. *Journal of Cleaner Production*, 248, 119182.

Rabbani, M., Heidari, R., and Yazdanparast, R. (2019). A stochastic multi-period industrial hazardous waste location-routing problem: Integrating NSGA-II and Monte Carlo simulation. *European Journal of Operational Research*, 272(3), 945–961.

Retnaraj, J., Selvan, S., Durairaj, J., Christyraj, S., Adhimoorthy, P., Rajagopalan, K., and Jebaranjitham, J.N. (2021). The Impact of the COVID-19 pandemic on green societies. In: Chakraborty, C., Roy, S., Sharma, S., Tran, T.A. (Eds.), *The Impact of the COVID-19 Pandemic on Green Societies*. Springer International Publishing, Berlin.

Roger, B., James, E.S., and David, D. (1989). *The Safe Disposal of Hazardous Wastes: The Special Needs and Problems of Developing Countries*. World Bank Technical Paper, 1(93), 154.

Roy, H., Rahman, T.U., Suhan, M.B., Al-Mamun, M.R., Haque, S., and Islam, M.S. (2022). A comprehensive review on hazardous aspects and management strategies of electronic waste: Bangladesh perspectives. *Heliyon*, 8(7).

Rucevska, I., Nellemann, C., Isarin, N., *et al.* (2015). *Waste Crime – Waste Risks Gaps in Meeting the Global Waste Challenge*. UNEP Rapid Response Assessment. United Nations Environment Programme and GRID-Arendal, Nairobi and Arendal. http://www.grida.no.

Sahiledengle, B. (2019). Self-reported healthcare waste segregation practice and its correlate among healthcare workers in hospitals of Southeast Ethiopia. *BMC Health Services Research*, 19(1), 591.

Samant, M., Pandey, S.C., and Pandey, A. (2018). Impact of hazardous waste material on environment and their management strategies. In: *Microbial Biotechnology in Environmental Monitoring and Cleanup*. IGI Global, pp. 175–192.

Sefouhi, L., Kalla, M., Bahmed, L., and Aouragh, L. (2013). The risk assessment for the healthcare waste in the hospital of Batna city, Algeria. *International Journal of Environmental Science and Development*, 4(4), 442.

Sharma, H.D. and Lewis, S.P. (1994). *Waste Containment Systems, Waste Stabilization, and Landfills: Design and Evaluation.* John Wiley and Sons, New York.

Sittig, M. (1991). *Handbook of Toxic and Hazardous Chemicals* (3rd ed.). Noyes Publications, Park Ridge, NJ.

Slack, R.J., Gronow, J.R., and Voulvoulis, N. (2005). Household hazardous waste in municipal landfills: contaminants in leachate. *Science of the Total Environment*, 337, 119–137.

Sun, S., Ren, J., Liu, J., Rong, L., Wang, H., Xiao, Y., Sun, F., Mei, R., Chen, C., and Su, X. (2022). Pyrite-activated persulfate oxidation and biological denitrification for effluent of biological landfill leachate treatment system. *Journal of Environmental Management*, 304, 114290.

Talyan, V., Dahiya, R.P., and Sreekrishnan, T.R. (2008). State of municipal solid waste management in Delhi, the capital of India. *Waste Management*, 28(7), 1276–1287.

Tejaswini, M.S.S.R., Pathak, P., Singh, L., Raj, D., and Gupta, D.K. (2023). A novel circular approach to analyze the challenges associated with micro-nano plastics and their sustainable remediation techniques. *Journal of Environmental Science and Health, Part A*, 58, 694–705.

The World Counts. (2023). Hazardous Waste Statistics. https://www.theworldcounts.com/challenges/planet-earth/waste/hazardous-waste-statistics. Accessed 6 July 2022.

UNEP. (2012). *The Global Garbage Crisis: No Time to Waste*. https://www.unep.org/news-and-stories/press-release/global-garbage-crisis-no-time-waste.

UNEP. (2020). *Solid Waste Management*. United Nations Environment Programme. https://www.unep.org/explore-topics/resource-efficiency/what-we-do/cities/solid-waste-management. Accessed 20 October 2021.

USEPA. (2005). *Municipal Solid Waste – Household Hazardous Waste*. Office of Solid Waste, US Environmental Protection Agency, Washington D.C.

USEPA. (2014). *Technical Fact Sheet – 2,4,6-trinitrotoluene (TNT)*. US Environmental Protection Agency, Washington, DC, pp. 1–8.

Via, S.M. and Zinnert, J.C. (2016). Impacts of explosive compounds on vegetation: A need for community scale investigations. *Environmental Pollution*, 208(Pt B), 495–505.

Walther, C. and Gupta, D.K. (2015). *Radionuclides in the Environment: Influence of Chemical Speciation and Plant Uptake on Radionuclide Migration.* Springer, Heidelberg.

Ward, E. (2011). Landfills a history. *Green Risks*. https://greenrisks.blogspot.com/2011/07/landfills-history.html?m=0. Accessed 12 July 2022.

Wath, S.B., Dutt, P.S., and Chakrabarti, T. (2011). E-waste scenario in India, its management and implications. *Environmental Monitoring and Assessment*, 172, 249–262

Weidenhamer, J.D., Newman, B.E., and Clever, A. (2010). Assessment of leaching potential of highly leaded jewelry. *Journal of Hazardous Materials*, 177, 1150–1152.

Wessel, J. (2004). *Household Hazardous Materials Programs, a Report to the Iowa General Assembly*. Iowa Department of Natural Resources, Energy and Waste Management Bureau, Iowa.

White, L. and Lee, G.J. (2009). Operational research and sustainable development: Tackling the social dimension. *European Journal of Operational Research*, 193(3), 683–692.

Widmer, R., Oswald-Krapf, H., Sinha-Khetriwal, D., Schnellmann, M., and Böni, H. (2005). Global perspectives on e-waste. *Environmental Impact Assessment Review*, 25(5), 436–458.

World Health Organization. (2011). *Wastes from Health-Care Activities*. Factsheet No. 253. http://www.who.int/mediacentre/factsheets/fs253/en.

Zabihian-Bisheh, A., Vandchali, H.R., Kayvanfar, V., and Werner, F. (2024). A sustainable multi-objective model for the hazardous waste location-routing problem: A real case study. *Sustainable Operations and Computers*, 5, 1–14.

Zhang, J., Wang, M., and Zhu, X. (2013). Treatment of HMX-production wastewater in an aerobic granular reactor. *Water Environment Research*, 85(4), 301–307.

Zhang, X., Gu, P., and Liu, Y. (2019). Decontamination of radioactive wastewater: State of the art and challenges forward. *Chemosphere*, 215, 543–553.

Zografos, K. and Samara, S. (1989). A combined location-routing model for hazardous waste transportation and disposal. *Transportation Research Record*, 1245, 52–59.

https://doi.org/10.1142/9789819822249_0002

Chapter 2

Understanding and Estimation of Emerging Contaminants in Soil: Sources, Effects and Impacts

Pankaj Pathak[*,§], **Rajiv R. Srivastava**[†,‡], **and Sadia Ilyas**[†,‡]

[*]*Resource Management Lab, Department of Environmental Science & Engineering, SRM University-AP, Andhra Pradesh 522240, India*

[†]*Process Metallurgy, Minerals and Metallurgical Engineering Division, Department of Civil, Environmental and Natural Resources Engineering, Luleå University of Technology, Luleå 97187, Sweden*

[‡]*Wallenberg Initiative Materials Science for Sustainability (WISE), Department of Civil, Environmental and Natural Resources Engineering, Luleå University of Technology, Luleå 97187, Sweden*

[§]*pankajpathak18@gmail.com*

Abstract

Globally, emerging contaminants (ECs) get huge attention, due to their negative impacts on the environment and human being, and necessitate strict regulatory guidelines. ECs include synthetic polymers, pharmaceuticals, personal care products, micro-nanoplastics, plasticizers, and flame retardants. The EC enters into the soil through various sources, viz., agricultural activities, household waste, and wastewater, etc., and spreads adverse impacts into the soil environment. It is noted that each EC behaves differently in the soil environment due to different characteristics of soil, such as soil type, organic content, and moisture. This chapter discusses the different sources of ECs, their effects, and impacts on the environment and human health.

Keywords: Emerging contaminants, environmental impacts, human impacts, soil.

1. Introduction

Emerging contaminants (ECs) show a global concern and are studied widely to determine their effects on the environment. ECs are first studied in water and then soil ecosystem (Tijani *et al.*, 2016). This emerging contaminant comprises almost 3000 types of compounds such as persistent organic pollutants, per-fluoroalkyl and poly-fluoroalkyl substances (PFASs), endocrine disruptors (ECDs), additives, flame retardants (FRs), micro-nano plastics (MNPs), and antibiotics that come out due to different applications, such as pesticides, pharmaceuticals, personal care products, disinfectants, surfactants, household materials, nanomaterials, and illicit drugs, refer to Table 1 (Puri *et al.*, 2023). It is reported that concentration of ECs is increasing in soil and sediment ecosystems due to leaching from the urban, industrial, and agricultural waste materials (Zhang *et al.*, 2024) and creates problems to the environment and human ecosystems (Liu *et al.*, 2024). Though the existence of ECs in the environment is not inevitably new, now, they have gained attention due to their potential negative impacts or are newly produced contaminants in the environment where soil system has become a potential sink, primarily due to agriculture activities, sewage sludge, etc. (Maddela *et al.*, 2022). However, there are no strict norms under existing regulations to assess and avoid such ECs from the environment, and this shows emerging concern. Though several research works have been done on assessing the mobility of ECs in the water systems, such as ground and surface water, wastewater, and the marine system of Asia, Africa, Australia, Europe, America, etc. (K'oreje *et al.*, 2020; Kroon *et al.*, 2020). However, less attention has been given

Table 1. Different types of emerging contaminants (ECs) and their sources.

ECs	**Sources**	**References**
Polybrominated diphenylethers	Electrical and electronic devices, textiles, plastics, packaged food products, and baby products	Klinčić *et al.* (2020)
Organophosphorus flame retardants	Polymers used in polyurethane foam, furniture, baby products, etc.	Breivik *et al.* (2016)
Micro-/ nano-plastics	Plastic debris	Amorim and Scott-Fordsmand (2021) and Tejaswini *et al.* (2023)
Additives–phthalates, biphenyls	Plastic additives, dielectric fluids, additives in lubricant fluids, adhesives, building sealants, plasticizers, fireproofing agents, paints, and ink products	Benjamin *et al.* (2015) and Breivik *et al.* (2016)
Per-and poly-fluoroalkyl substances (PFASs)	Food package material, stain-and water-repellent fabrics, non-stick products (e.g., Teflon), polishes, waxes, paints, cleaning products, fire-fighting foams, industrial facilities (e.g., chrome plating, electronic goods, and oil recovery), landfill wastewater treatment plant, and living organisms (e.g., fish, animals, and humans) due to accumulation and persistence over time	USEPA (2018) and Li *et al.* (2023)
Nanoparticles	Cosmetics, pharmaceuticals, coatings, paints, and catalytic additives	Nandikes *et al.* (2024)

Source: Adopted from Maddela *et al.* (2022).

to the soil pollution due to ECs, and therefore, tremendous burden is on soil system globally (Maddela *et al.*, 2022).

Additionally, due to an advanced and sensitive analytical technique, it is possible to estimate the concentration of ECs and show their presence in the soil, air, and water even at a very low concentration level as parts per trillion (PPT). Ion exchange, adsorption, sorption/desorption,

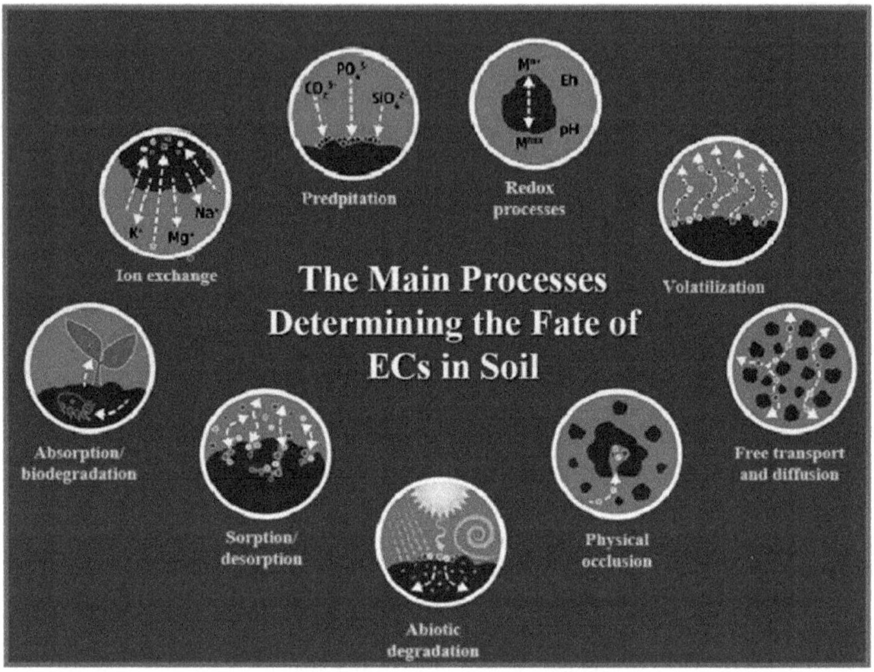

Fig. 1. Different processes to determine the fate of ECs in soil.
Source: Adopted from Liu *et al.* (2024).

abiotic degradation, physical occlusion, etc. are different methods used in assessing the fate of ECs in the soil system, as shown in Fig. 1 (Liu *et al.*, 2024).

Therefore, it is utmost important to identify the sources of ECs, impacts associated with their response to soil and soil microbiome, and the bioaccumulation and biomagnification of ECs in the environment. Major ECs, such as flame retardants, PFASs, micro-nano plastics, and additives, are discussed in this chapter.

2. Types of Different ECs in the Soil

The ECs either exist in the environment and due to their increasing concentration are brought into the emerging concern category or are newly produced with the pollutants and have severe impacts (Tejaswini *et al.*, 2023). Four major ECs are selected and discussed in the following.

2.1. *Polybrominated diphenyl ethers: flame retardants*

Globally, the polybrominated diphenyl ethers (PBDEs) and hexabromocyclododecane (HBCDD) are produced under a class of brominated flame retardants and widely present in abiotic and biotic environment, including humans (Klinčić *et al.*, 2020). Theoretically, there are around 209 possible PBDE congeners that are divided into 10 congener groups based on the number of bromine atoms, but all share a common structure entailing a brominated diphenyl ether molecule with two benzene rings coupled with an oxygen atom, as shown in Fig. 2 (Klinčić *et al.*, 2020; La Guardia *et al.*, 2006). The PBDEs and HBCDD compounds are persistent in nature for longer and show bioaccumulation in the living system. Due to this, it can reach the human being and disrupt the human's endocrine system, nervous system, and reproductive system, etc. (Chen *et al.*, 2019).

2.2. *Per- and polyfluoroalkyl substances (PFASs)*

The per- and polyfluoroalkyl substances (PFASs) are resistant to water, oil, stains, grease, and heat, which have fluoropolymer coatings as a major component with a class of over 12,000 chemicals (USEPA, 2024). However, due to strong chemical bonds of PFASs, they remain longer in the environment and do not get degraded easily (Salvatore *et al.*, 2022). Therefore, it is widely used in different industrial and consumer products, such as waterproof coatings, food packaging, non-stick cooking surfaces, electric wire insulation, and household items. PFASs are also used in Class B fire-fighting foam due to their heat-stable properties. Two types of PFASs are commonly used: perfluorooctanesulfonic acid (PFOS) and perfluorooctanoic acid (PFOA). Usually, PFASs are hydrophobic, fluorine-alkylated compounds that contain saturated carbon chains and

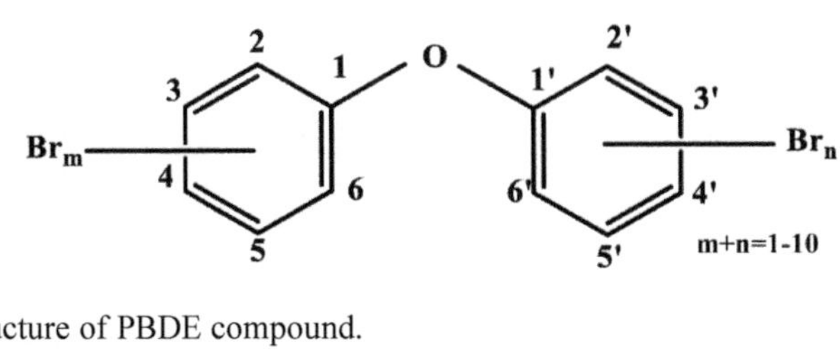

Fig. 2. Structure of PBDE compound.

Sources: Klinčić *et al.* (2020) and La Guardia *et al.* (2006).

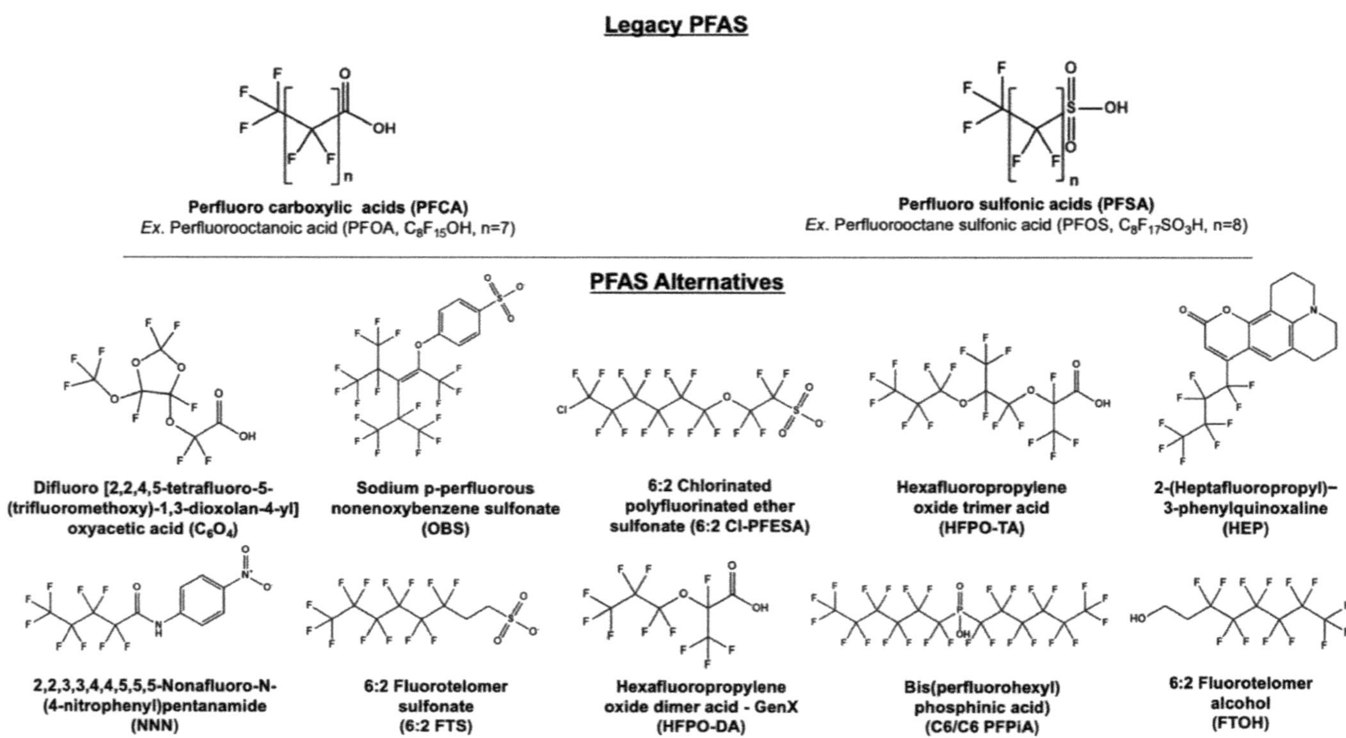

Fig. 3. Legacy and alternative of PFAS.

Source: Adopted from Dickman and Aga (2022).

have a hydrophilic head (sulfonate or carboxylate) attached at a terminal end (Li *et al.*, 2023). Further, new PFAS substitutes were produced through their structural modifications to make them easily degradable. These PFAS alternatives have ether or chlorinated carbons linked with shorter fluoroalkyl chains, as shown in Fig. 3 (Dickman and Aga, 2022).

Nevertheless, it is noted that PFAS has an intractable nature and is able to transfer into different environments such as soil and water and show bioaccumulation (USEPA, 2024).

2.3. *Micro-nano plastics (MNPs)*

Micro and nano plastics (MNPs) having sizes micro and nano level, i.e., $\leq$ 5.0 mm and $\geq$1.0 nm, are one of the prominent ECs worldwide and show a serious concern. Primary and secondary sources are responsible for entering the MNPs into the environment where primary sources include discarded plastic wastes from clothing fibers, cosmetics from urban system, and mulches and polytunnels in the agroecosystem (Tejaswini *et al.*, 2023). It has been noted that each wash of personal care products can produce almost 94,500 microbeads (synthesized solid plastic particles) per day. Whereas, fragmentation of primary plastics products in the presence of UV/visible radiation forms the MNPs and is known as secondary contaminant (Jones and Pathak, 2025; Kumar *et al.*, 2024). Surprisingly, almost 80% of MNPs are accumulated in the landfill as compared to their other environment, such as ground and surface water and ocean (Nandikes *et al.*, 2024). The intensity of MNP contaminants is severe in the agricultural soil as compared to oceanic basin (Wang *et al.*, 2022). MNPs' response on soil ecosystem is different and able to alter their properties, which shows detrimental impacts on plants and animals, reported by Zhang *et al.* (2022), refer Fig. 4. The MNPs are long-lasting with very less biodegradability and get accumulated rapidly in the environment (Jones and Pathak, 2025), and show abundance of six orders of magnitude with different terrestrial environments (Koutnik *et al.*, 2021). It is reported that around 63,000–430,000 and 44,000–300,000 tons of MNPs have been produced annually in European and North American farmland soils, respectively (Ng *et al.*, 2018). In Korea and Canada, MNP concentrations in agricultural soils have been found to be about 10–7630 pieces.kg^{-1} and 4–541 pieces.kg^{-1} of soil, respectively (Crossman *et al.*, 2020).

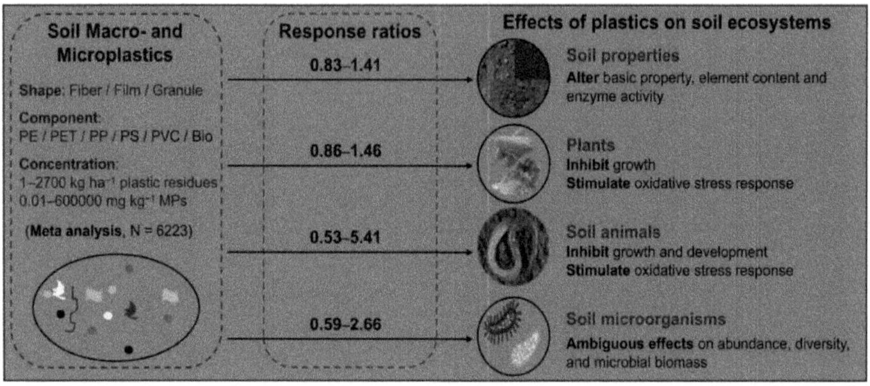

Fig. 4. MNPs in soil ecosystems and their effects.
Source: Adopted from Zhang *et al.* (2022).

2.4. *Biphenyls/phthalates: additives*

The biphenyls and phthalate esters are common organic contaminants in soil because of their applications in plastic products, pesticides, cosmetics, personal care products, building materials, automotive parts, food packaging, toys, teethers, adhesives, paints, and lubricants. It is observed that plastic industry has used more than 3 million metric tons of phthalates per year and global production of phthalate has almost doubled from 4.7 to 8.0 million metric tons between 2006 and 2015 (Rushdi *et al.*, 2023). These organic contaminants leached out into the soil and transfer to plants and animals and cause severe impacts.

3. Environmental and Health Impacts Associated with ECs

The ECs have created a mounting curiosity to understand their intricate interactions with the soil ecosystem. Different types of microorganisms and plants are present in the soil that accumulate ECs through several processes, such as adsorption, ingestion, metabolism, and transport. Through these processes, EC slowly enters the food chain and becomes part of it (Ma *et al.*, 2024). Accumulation of ECs in the soil is highly dependent on soil properties; either ECs will get transferred to the plants and crops, or they will leach out from the soil and enter into groundwater

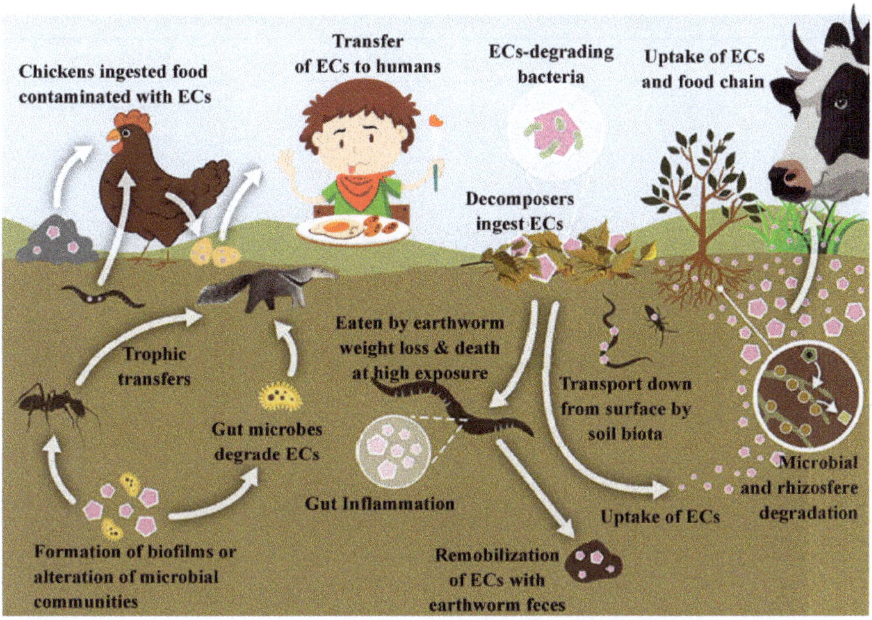

Fig. 5. Bioaccumulation and biomagnification of ECs into the terrestrial food web from the soil system.

Source: Adopted from Liu *et al.* (2024).

(Jones and Pathak, 2025). It is demonstrated through Fig. 5 that ECs are consumed by microbes or accumulated by plants and crops. And further, they are transferred from one trophic level to another trophic level. During this process, the concentration of ECs gets increased from lower to higher trophic levels and shows biomagnification, refer Fig. 5 (Ma *et al.*, 2024). Therefore, it is of utmost significance to determine the potential hazards associated with ECs before they reach the soil system.

One of the prominent ECs is brominated flame retardants, such as PBDE and HBCDD, that found in the soil, animals, and human being. Globally, the concentration of HBCDD is increasing and its impacts have been seen in humans where it has the ability to disrupt the endocrine system by mirroring the estrogen/thyroid/androgen receptors. Generally, humans are exposed to FRs such as PBDEs through diet, inhalation, ingestion of dust, and dermal contact. It has been studied that the principal source of FR exposure is food of animal origining with greater fat content (fish, meat, and dairy products). In this context, Ma *et al.* (2024) have

Table 2. Concentrations of HBCDD (in pg/g ww) in food items from different countries.

Country	Analysis Year	Sample Type	HBCDD (pg/g)	References
UK	2020–2021	Meat	0.16	Ma *et al.* (2023)
		Fish	0.83	
		Eggs	0.19	
		Cheese	0.050	
Latvia	2016–2019	Meat	270	Zacs *et al.* (2021)
		Fish	730	
		Eggs	24	
		Cheese	330	
China	2013	Meat	2.8	Wang *et al.* (2019a)
		Fish/seafood	7.0	
		Eggs	4.5	
		Dairy products	1.4	
France	2014–2016	Meat	130	Venisseau *et al.* (2018)
		Fish	140	
Belgium	2015–2016	Meat	43	Poma *et al.* (2018)
		Fish	130	
		Eggs	970	
		Dairy products	0.10	

Source: Adopted from Ma *et al.* (2024).

summarized the concentration of HBCDD in food items and listed them in Table 2.

Another EC which has severe concerns in the environment is PFAS, whose concentration has increased tremendously. Strynar *et al.* (2012) have reported that globally around 1860–7000 metric tons of PFASs are loaded in the soil environment and the maximum concentrations of PFOA and PFOS (combining PFAS) were 124 $mg.kg^{-1}$ and 162 $mg.kg^{-1}$, respectively, in the soil. It has been found that crops' root accumulates PFAS in a faster way and reveals the major accumulation pathway in the food crops that shows a significant menace to food safety and human health, refer Fig. 6 (Wang *et al.*, 2020).

Additionally, PFAS shows significant impacts on animals and disrupts endocrine activities. It is also studied that it induced the morphological

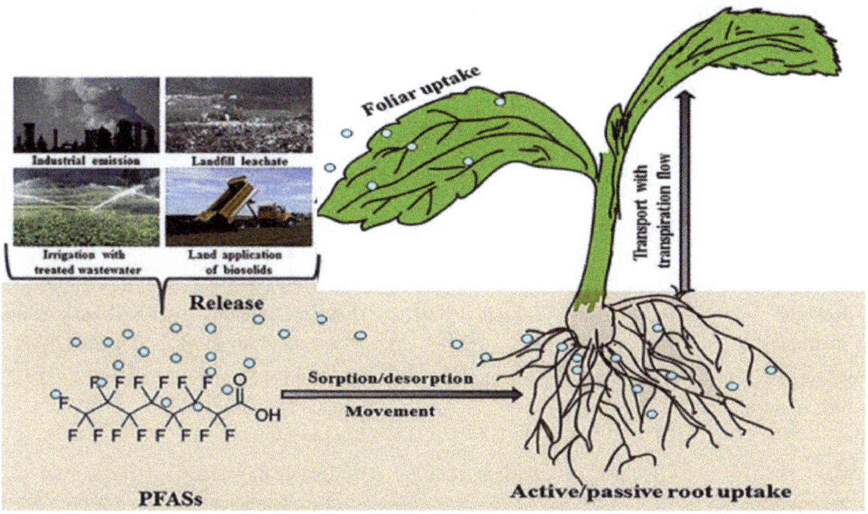

Fig. 6. Uptake capacity of PFAS in soil and crop root system.

Source: Adopted from Wang *et al*. (2020).

and functional perturbations during pancreatic organogenesis in zebrafish (Sant *et al*., 2021); however, in human, it tends to interrupt the production of human pancreatic progenitor cells and interfere with pancreatic and endocrine division (Kato *et al*., 2014). It has been found that different classes of PFASs are responsible for the ecological, animal, and human toxicities.

Furthermore, MNPs are also known as secondary pollutants and cause severe impacts on environment and human being. It has been noted that MNPs are accumulated in soil and greatly affect the soil properties. Mostly, MNPs have a hydrophobic surface; due to this, it may alter the water holding capacity, transport, and availability in soil system and is able to change the soil fertility and contaminant toxicity (Jones and Pathak, 2025; Wang *et al*., 2022). The major contributors of MNPs in agricultural soil are mulching and biosolids that come out from wastewater treatment plants. MNPs reach soil biota and are consumed by their organisms and their effects are shown in Table 3.

Thus, MNPs get transferred through the food chain and pose an impending risk to human health. It has been reported that MNPs affect the human immune cells by different routes such as phagocytosis and micropinocytosis, and this interaction between immune cells and MNPs can

Table 3. Effects of MNPs on different types of soil organisms.

MPs	Size	Species	Effects	Reference
Polyurethane foam microparticles	3.9–33.4 μm	*Eisenia fetida*	Toxic accumulation and transformation to predators	Gaylor *et al.* (2013)
Low-density polyethylene (LDPE) microplastics	< 150 μm	*Lumbricus terrestris*	Toxic effect on organism indicated from growth and mortality	Lwanga *et al.* (2017)
Polyethene microplastics	250–1000 μm	*Eisenia andrei*	Damaged immune system of organism	Rodriguez-Seijo *et al.* (2017)
Polystyrene microplastics	1–5 μm	Nematodes	Damaged cholinergic and GABAergic neurons	Lei *et al.* (2018)
PVC microplastic	80–250 μm	Collembolan	Restrained movement of Collembolan	Kim and An (2019
Polyethylene microplastics	183–93 μm	Isopods	No obvious effect	Kokalj *et al.* (2018)
LDPE	< 400 μm particles	Earthworm, *Lumbricus terrestris*	Mortality	Huerta *et al.* (2016)
PET	1257.8 μm fiber	Snails, *Achatina fulica*	Affected oxidative stress	Song *et al.* (2017)
PS	0.1–500 μm residues	*Caenorhabditis elegans*	Oxidative damage	Lei *et al.* (2018)
LDPE	≤ 300 μm particles	Earthworm, *Eisenia fetida*	Microplastic increased addition of PAHs in addition to PCBs in *E. fetida*.	Wang *et al.* (2019b)

Source: Adopted from Sajjad *et al.* (2022).

induce severe immune dysfunctions (Sajjad *et al.*, 2022; Zhang *et al.*, 2022).

Another selected ECs which are phthalate esters (PAEs) are commonly used as plasticizers in electronic devices and strongly adsorb on the soil surface. Therefore, the concentration of PAEs is higher on the surface, and as we go into deeper layers of the soil, PAEs have lesser concentration (Tran *et al.*, 2022). PAEs generally reach human life through ingestion of contaminated food, inhalation of indoor air, contaminated dust particles, and dermal application of personal care products. However, PAEs belong to the group of chemicals which disrupt the endocrine activities even at low concentrations. It is reported that this phthalate ester can affect androgen antagonism and decrease thyroid hormones (Benjamin *et al.*, 2017).

4. Environmental Regulation on ECs

The ECs have become a critical soil pollutant because of their environmental persistence and mobility. It is not feasible to depend only on natural attenuation (EC degradation without human involvement) or long-term monitoring to restore EC-contaminated soil, as the concentration of ECs is increasing in the environment. Due to that, it has the ability to easily leach out from the soil and enter the aquatic system, i.e., groundwater, marine system. However, there is no strict regulation made by any country to treat or manage the ECs. But few significant regulatory frameworks have been introduced that support research on ECs in the environment. In this context, ECs were first reported in the aquatic environment, and serious concern was made by the US Geological Survey. Around 139 streams were analyzed around in the USA and determined the presence of pharmaceutical and personal care items up to 80%. Based on this survey, the US Environmental Protection Agency (EPA) made a mandate on health advisory levels based on ECs. In addition, the Toxic Substances Control Act (TSCA) and the Food and Drug Administration (FDA) have classified the hazards associated with and suggested to make laws to regulate the ECs in the environment (Puri *et al.*, 2023). In this context, USEPA has made regulations for permissible limit for PFASs as 70 parts per trillion (ppt). Nevertheless, the New Jersey Department of Environmental Protection (NJDEP) has shown the values of 14 and 13 ppt for PFOA and PFOS, respectively. The Australian PFAS National Environmental Management Plan (NEMP) has made the PFAS toxicity level for soil in the range of 0.01–50.0 $mg.kg^{-1}$. However, there is still a need to frame strict action plans, regulations, and risk control rules, along with technical guidelines for PFAS pollution in soil.

Another global concern is given to MNPs, where different regulations are framed to remove MNPs from the environment. Interestingly, several countries such as the Netherlands, the USA, Canada, Australia, and India have made laws to ban single-use plastics, and microbeads in cosmetics have minimized the load of MNPs in soil and sewage sludge. For instance, the aim of 'The EU strategy for plastics' and 'Plastic Waste management Rule in India' states that plastic packaging is to be reusable and recyclable by 2030. Further, in Germany, the permissible quantity of plastics in fertilizer is 0.1% weight, but <2 mm size is not considered. On the other hand, EU fertilizer legislation allows 0.5% weight of plastics in organic fertilizers. However, there is no specific rule, and regulations on MNPs that directly provide guidelines to remove them from the environment.

5. Conclusions

In conclusion, the widespread presence of ECs in soils presents significant risks to both environmental health and human well-being. These chemicals, including brominated flame retardants, PFASs, microplastics, and phthalate esters, accumulate in soil and transfer through the food chain, leading to biomagnification. Despite some regulatory efforts, such as the US EPA's guidelines, existing regulations remain inadequate. To address the growing threat of ECs, more stringent laws, enhanced risk management strategies, and continued research are critical. Given the emerging concerns about their genotoxic and carcinogenic potential, it is essential to critically analyze their impact on soil, transfer pathways, and formulate robust policies to safeguard public health and ecosystems.

Acknowledgments

This work was partially supported by the Wallenberg Initiative Materials Science for Sustainability (WISE) funded by the Knut and Alice Wallenberg Foundation.

Bibliography

Amorim, M.J.B. and Scott-Fordsmand, J.J. (2021). Plastic pollution – A case study with Enchytraeuscrypticus – From micro-to nanoplastics. *Environmental Pollution*, 271, 116363.

Benjamin, S., Pradeep, S., Josh, M.S., Kumar, S., and Masai, E. (2015). A monograph on the remediation of hazardous phthalates. *Journal of Hazardous Materials*, 298, 58–72.

Benjamin, S., Masai, E., Kamimura, N., Takahashi, K., Anderson, R.C., and Faisal, P.A. (2017). Phthalates impact human health: Epidemiological evidences and plausible mechanism of action. *Journal of Hazardous Materials*, 340, 360–383.

Breivik, K., Armitage, J.M., Wania, F., Sweetman, A.J., and Jones, K.C. (2016). Tracking the global distribution of persistent organic pollutants accounting for e-waste exports to developing regions. *Environmental Science & Technology*, 50, 798–805.

Chen, T., Huanga, M., Li, J., *et al.* (2019). Polybrominated diphenyl ethers and novel brominated flame retardants in human milk from the general population in Beijing, China: Occurrence, temporal trends, nursing infants' exposure and risk assessment. *Science of the Total Environment*, 689, 278–286.

Crossman, J., Hurley, R.R., Futter, M., and Nizzetto, L. (2020). Transfer and transport of microplastics from biosolids to agricultural soils and the wider environment. *Science of the Total Environment*, 724, 138334.

Dickman, R.A. and Aga, D.S. (2022). A review of recent studies on toxicity, sequestration, and degradation of per- and polyfluoroalkyl substances (PFAS). *Journal of Hazardous Materials*, 436, 129120.

Gaylor, M.O., Harvey, E., and Hale, R.C. (2013). Polybrominated diphenyl ether (PBDE) accumulation by earthworms (Eisenia fetida) exposed to biosolids-, poly urethane foam microparticle-, and Penta-BDE-amended soils. *Environmental Science & Technology*, 47, 13831–13839.

Huerta, L.E., Gertsen, H., Gooren, H., Peters, P., Salanki, T., van der Ploeg, M., *et al.* (2016). Microplastics in the terrestrial ecosystem: implications for Lumbricusterrestris (Oligochaeta, Lumbricidae). *Environmental Science & Technology*, 50(5), 2685–2691.

Jones, K.H. and Pathak, P. (2025). Recycling technologies and resource recovery from plastic waste. In: *Valorization of Resources from Urban Mined Materials*. World Scientific Series on Advances in Environmental Pollution Management, World Scientific, pp. 61–80.

Kato, K., Wong, L.Y., Chen, A., Dunbar, C., Webster, G.M., Lanphear, B.P., and Calafat, A.M. (2014). Changes in serum concentrations of maternal poly- and perfluoroalkyl substances over the course of pregnancy and predictors of exposure in a multiethnic cohort of Cincinnati, Ohio pregnant women during 2003–2006. *Environmental Science & Technology*, 48(16), 9600–9608.

Kim, S.W. and An, Y.J. (2019). Soil microplastics inhibit the movement of spring tail species. *Environment International*, 126, 699–706.

Klinčić, D., Dvoršćak, M., Jagić, K., Mendaš, G., and Romanić, S.H. (2020). Levels and distribution of polybrominated diphenyl ethers in humans and environmental compartments: A comprehensive review of the last five years of research. *Environmental Science and Pollution Research*, 27, 5744–5758.

Kokalj, A.J., Horvat, P., Skalar, T., and Krzan, A. (2018). Plastic bag and facial cleanser derived microplastic do not affect feeding behaviour and energy reserves of terrestrial isopods. *Science of the Total Environment*, 615, 761–766.

K'oreje, K.O., Okoth, M., Langenhove, H.V., and Demeestere, K. (2020). Occurrence and treatment of contaminants of emerging concern in the African aquatic environment: Literature review and a look ahead. *Journal of Environmental Management*, 254, 109752.

Koutnik, V.S., Leonard, J., Alkidim, S., DePrima, F.J., Ravi, S., *et al.* (2021). Distribution of microplastics in soil and freshwater environments: Global analysis and framework for transport modelling. *Environmental Pollution*, 274, 116552.

Kroon, F.J., Berry, K.L.E., Brinkman, D.L., Kookana, R., Leusch, F.D.L., *et al.* (2020). Sources, presence and potential effects of contaminants of emerging concern in the marine environments of the Great Barrier Reef and Torres Strait, Australia. *Science of the Total Environment*, 719, 135140.

Kumar, P., Jones, H., Pathak, P., and Singh, P. (2024). Automated segregation of plastic wastes: Pioneering a greener frontier in waste management. In: *Advances in Environmental Sustainability, Energy and Earth Science. AESEE 2024*. Springer, Cham. https://doi.org/10.1007/978-3-031-73820-3_1.

La Guardia, M.J., Hale, R.C., and Harvey, E. (2006). Detailed polybrominated diphenyl ether (PBDE) congener composition of the widely used penta-, octa-, and deca-PBDE technical flame-retardant mixtures. *Environmental Science & Technology*, 40(20), 6247–6254.

Lei, L., Liu, M., Song, Y., Lu, S., Hu, J., Cao, C., Xie, B., Shi, H., and He, D. (2018). Polystyrene (nano) microplastics cause size-dependent neurotoxicity, oxidative damage and other ad verse effects in caenorhabditis elegans. *Environmental Science: Nano*, 5, 2009–2020.

Li, H., Dong, Q., Zhang, M., Gong, T., Zan, R., and Wang, W. (2023). Transport behavior difference and transport model of long- and short-chain per- and polyfluoroalkyl substances in underground environmental media: A review. *Environmental Pollution*, 327, 121579.

Liu, L., Liu, C., Fu, R., Nie, F., Zuo, W., Tian, Y., and Zhang, J. (2024). Full-chain analysis on emerging contaminants in soil: Source, migration and remediation. *Chemosphere*, 363, 142854.

Lwanga, E.H., Gertsen, H., Gooren, H., Peters, P., Salanki, T., van der Ploeg, M., Besseling, E., Koelmans, A.A., and Geissen, V. (2017). Incorporation of microplastics from litter into burrows of Lumbricusterrestris. *Environmental Pollution*, 220, 523–531.

Ma, Y., Stubbings, W.A., Abdallah, M.A.E., Cline-Cole, R., and Harrad, S. (2023). Temporal trends in concentrations of brominated flame retardants in UK foodstuffs suggest active impacts of global phase-out of PBDEs and HBCDD. *Science of the Total Environment*, 863, 160956.

Ma, Y., Stubbings, W.A., Cline-Cole, R., Abdallah, M.A., and Harrad, S. (2024). Rising concentrations of polybrominated diphenyl ethers (PBDEs) in Nigerian foodstuffs despite global restrictions. *Emerging Contaminants*, 10(1), 100264.

Maddela, N.R., Ramakrishnan, B., Kakarla, D., Venkateswarlue, K., and Megharaj, M. (2022). Major contaminants of emerging concern in soils: A perspective on potential health risks. *RSC Advances*, 12, 12396.

Nandikes, G., Banerjee, O., Mirthipati, M., Bhargavi, A., Jones, H., and Pathak, P. (2024). Separation, identification, and quantification of microplastics in environmental samples. *ACS Symposium Series*. American Chemical Society, Vol. 1482, pp. 1–19.

Ng, E.L., Lwanga, E.H., Eldridge, S.M., Johnston, P., *et al.* (2018). An overview of microplastic and nanoplastic pollution in agroecosystems. *Science of the Total Environment*, 627, 1377–1388.

Poma, G., Malysheva, S.V., Goscinny, S., Malarvannan, G., Voorspoels, S., Covaci, A., and Loco, J.V. (2018). Occurrence of selected halogenated flame retardants in Belgian foodstuff. *Chemosphere*, 194, 256–265.

Puri, M., Gandhi, K., and Kumar, M.S. (2023). Emerging environmental contaminants: A global perspective on policies and regulations. *Journal of Environmental Management*, 332, 117344.

Rodriguez-Seijo, A., Lourenço, T.A.P., Costa, J.D., Duarte, A.C., Vala, H., and Pereira, R. (2017). Histopathological and molecular effects of micro plastics in Eisenia andrei Bouche. *Environmental Pollution*, 220, 495–503.

Rushdi, A.I., Simoneit, B.R.T., Lijotra, L., Bazeyad, A.Y., Dumenden, R., El-Mubarak, A.H., Qurban, M.A., and Al-Mutlaq, K.F. (2023). Phthalates, non-phthalates, polychlorinated biphenyls, and phenyl phosphates in atmospheric suspended particulate matter of Dhahran City, Saudi Arabia: Levels and seasonal variation. *International Journal of Environmental Science and Technology*, 20, 3561–3576.

Sajjad, M., Huang, Q., Khan, S., Khan, M.A., Liu, Y., *et al.* (2022). Microplastics in the soil environment: A critical review. *Environmental Technology & Innovation*, 27, 102408.

Salvatore, D., Mok, K., Garrett, K.K., Poudrier, G., Brown, P., *et al.* (2022). Presumptive contamination: A new approach to PFAS contamination based on likely sources. *Environmental Science & Technology Letters*, 9(11), 983–990.

Sant, K.E., Annunziato, K., Conlin, S., Teicher, G., Chen, P., Venezia, O., Downes, G.B., *et al.* (2021). Developmental exposures to perfluorooctanesulfonic acid (PFOS) impact embryonic nutrition, pancreatic morphology, and adiposity in the zebrafish, *Danio rerio*. *Environmental Pollution*, 275, 116644.

Song, Y.K., Hong, S.H., Jang, M., Han, G.M., Jung, S.W., and Shim, W.J. (2017). Combined effects of UV exposure duration and mechanical abrasion on microplastic fragmentation by polymer type. *Environmental Science & Technology*, 51, 4368–4376.

Strynar, M.J., Lindstrom, A.B., Nakayama, S.F., Egeghy, P.P., and Helfant, L.J. (2012). Pilot scale application of a method for the analysis of perfluorinated compounds in surface soils. *Chemosphere*, 86(3), 252–257.

Tejaswini, M.S.S.R., Pathak, P., Singh, L., Raj, D., and Gupta, D.K. (2023). A novel circular approach to analyze the challenges associated with micro-nano plastics and their sustainable remediation techniques. *Journal of Environmental Science and Health, Part A*, 58(7), 694–705.

Tijani, J.O., Fatoba, O.O., Babajide, O.O., and Petrik, L.F. (2016). Pharmaceuticals, endocrine disruptors, personal care products, nanomaterials and perfluorinated pollutants: A review. *Environmental Chemistry Letters*, 14, 27–49.

Tran, H.T., Lin, C., Bui, X.T., Nguyen, M.K., Cao, N.D.T., Mukhtar, H., Hoang, H.G., Varjani, S., Ngo, H.H., and Nghiem, L.D. (2022). Phthalates in the environment: Characteristics, fate and transport, and advanced wastewater treatment technologies. *Bioresource Technology*, 344, 126249.

United States Environmental Protection Agency (USEPA). (2018) PFOA, PFOS and Other PFASs. Basic Information on PFAS-What are PFAS? https://www.epa.gov/pfas/basic-information-pfas.

United States Environmental Protection Agency (USEPA). (2024). EPA's PFAS Strategic Roadmap: Three Years of Progress. https://www.epa.gov/system/files/documents/2024-11/epas-pfas-strategic-roadmap-2024_508.pdf. Accessed 25 January 2025.

Venisseau, A., Bichon, E., Brosseaud, A., Vaccher, V., Lesquin, E., Larvor, F., Durand, S., Dervilly-Pinel, G., *et al.* (2018). Occurence of legacy and novel brominated flame retardants in food and feed in France for the period 2014 to 2016. *Chemosphere*, 207, 497–506.

Wang, J., Zhao, X., Wang, Y., and Shi, Z. (2019a). Tetrabromobisphenol A, hexabromocyclododecane isomers and polybrominated diphenyl ethers in foodstuffs from Beijing, China: Contamination levels, dietary exposure and risk assessment. *Science of the Total Environment*, 666, 812–820.

Wang, W., Gao, H., Jin, S., Li, R., and Na, G. (2019b). The ecotoxicological effects of microplastics on aquatic food web, from primary producer to human: A review. *Ecotoxicology and Environmental Safety*, 173, 110–117.

Wang, W., Rhodes, G., Ge, J., Yu, X., and Li, H. (2020). Uptake and accumulation of per- and polyfluoroalkyl substances in plants. *Chemosphere*, 261, 127584.

Wang, F., Wang, Q., Adams, C.A., Sun, Y., and Zhang, S. (2022). Effects of microplastics on soil properties: Current knowledge and future perspectives. *Journal of Hazardous Materials*, 424, 127531.

Zacs, D., Perkons, I., Abdulajeva, E., Pasecnaja, E., Bartkiene, E., and Bartkevics, V. (2021). Polybrominated diphenyl ethers (PBDEs), hexabromocyclododecanes (HBCDD), dechlorane-related compounds (DRCs), and emerging brominated flame retardants (EBFRs) in foods: The levels, profiles, and dietary intake in Latvia. *Science of the Total Environment*, 752, 141996.

Zhang, J., Ren, S., Xu, W., Liang, C., Li, J., Zhang, H., Li, Y., *et al.* (2022). Effects of plastic residues and microplastics on soil ecosystems: A global meta-analysis. *Journal of Hazardous Materials*, 435, 129065. https://doi.org/10.1016/j.jhazmat.2022.129065.

Zhang, C., Liu, A., and Bahar, M. (2024). Microbial response to emerging contaminants in soil and sediment ecosystems. *Frontiers in Microbiology*, 15, 1371223. 10.3389/fmicb.2024.1371223.

https://doi.org/10.1142/9789819822249_0003

Chapter 3

Physical/Chemical Impact of Hazardous Waste on Soil

Monisha Mondol* **and Prodipto Bishnu Angon**[†,‡]

[*]Environmental Science Discipline, Life Science School, Khulna University, Khulna, 9208, Bangladesh

[†]Institute of Agronomy, Hungarian University of Agriculture and Life Science (MATE), Gödöllő, 2100, Hungary

[‡]angonbishnubau@gmail.com

Abstract

Soil is a vital ecosystem and a natural substrate for plant growth. Apart from providing a home for various organisms, it also participates directly or indirectly in several biogeochemical cycles. Neglecting garbage in the open seriously affects the environment, including decreased vegetation quantity, deteriorated soil quality, and soil microbes. In particular, soil fertility, crop growth performance, and soil microorganisms can all be significantly impacted by the physical and chemical impacts of hazardous waste (HW) materials because heavy metals (Pb, Hg, and Cd), persistent organic pollutants (POPs), hydrocarbons, and synthetic chemicals are among the harmful materials that are frequently included in HW. HW's physical and chemical effects on soil can cause long-term environmental harm, a decline in biodiversity, and major health hazards for people and animals. This chapter focuses on various sources of hazardous soil waste and the consequences of hazardous substances on human and soil health. We address the different physical and chemical effects of hazardous waste on soil habitat, identify research gaps, and outline opportunities for further study on the effects of hazardous waste on soil habitat.

Keywords: Soil microbes; Hazardous waste; Vegetation; Soil habitat; Physical-chemical effects.

1. Introduction

Soil is the foundation of agriculture. It is essential for all crops grown for human consumption and animal feed (Aqeel *et al.*, 2014). There are health risks and risks for soil composition associated with importing hazardous wastes (HWs) for recycling or disposal when a nation has few industries that produce these pollutants (Orloff and Falk, 2003). Any solid or liquid waste deemed toxic, chemically reactive, combustible, or corrosive is generally referred to as hazardous waste (HW). When inhaled, consumed, or touched, a chemical is detrimental to human health; it is classified as toxic in the context of HW (Shukla *et al.*, 2020). Furthermore, the massive amounts of sludge, other goods, and man-made waste products from new waste treatment plants, even contaminated water, are contributing to or creating soil pollution (Aqeel *et al.*, 2014). The definitions of HWs vary, making it challenging to compare the amounts of HW produced in various nations. Although significant amounts of hazardous garbage are burned in some nations, landfilling is the most popular method of disposing of HW in most nations. Humans are the most frequently impacted by HWs that leak into the environment because they contaminate the air, water, and soil (Orloff and Falk, 2003). Around 70% of all hazardous trash produced in the United States comes from the petroleum and chemical industries, which generate the most hazardous garbage (Shukla *et al.*, 2020). Public health emphasizes pressing issues, including infant mortality, hunger, and infectious diseases in underdeveloped nations. The health risks associated with hazardous chemical wastes are given more attention when a nation grows and acquires economic resources (Orloff and Falk, 2003).

Various contaminants and wastes pollute the soil. The accumulation of chemicals, salts, radioactive materials, persistent poisonous substances, or disease-causing agents in soils that negatively impact plant and animal health is known as soil pollution (Zhang *et al.*, 2012; Orloff and Falk, 2003). HWs, with some soil contamination from various causes, hinder land utilization in many developed nations (Van Straalen, 2002). Urban runoff, agricultural effluents from animal husbandry, industrial wastes containing various contaminants, and the uncontrolled disposal of sewage and other liquid wastes from household water use are all common causes of soil contamination (Aqeel *et al.*, 2014). Numerous contaminants from wastes are found in high concentrations in the blood, placentas, hair, and other bodily fluids of people who live near waste sites, endangering the population. These pollutants are also carried to the food chain through rice and other grains from the soil (Zhang *et al.*, 2012; Orloff and Falk, 2003).

HWs are a necessary byproduct of any industrialized society, but their detrimental impacts on the environment and human health are making them a major issue. Various HWs result in various health effects on soil and environmental health (Table 1). The most prevalent types of industrial hazardous waste often include cleaning solvents, wastewater from

Table 1. Various hazardous wastes with their effects on soil health.

Hazardous waste	Waste source	Effect on soil health	References
		Toxic chemicals	
Furans	Organic solvents	Degrade soil	Holtzer *et al.* (2016)
Toluene	Radioactive materials	Affects nutrient cycling and soil structure	Holtzer *et al.* (2016)
Ethylbenzene	Petroleum hydrocarbons	Alter structure and texture and reduce water holding capacity	Soares *et al.* (2011)
		Organic chemicals	
Dioxins	Fertilizers	Changes soil chemistry by binding organic and inorganic particles	Elskens *et al.* (2013)
PCBs	Historical contamination, synthetic organic compound	Alter soil water retention capacity	Zhang *et al.* (2012) and Zeeb *et al.* (2006)

(*Continued*)

Table 1. (*Continued*)

Hazardous waste	Waste source	Effect on soil health	References
		Heavy metal	
Lead (Pb)	Battery waste	Disrupt soil porosity and erode soil	Dixit and Srivastava (2015)
Chromium (Cr)	Steel or metal manufacturing, battery waste, and mining	Changes in color, structure, and chemical properties	Testa (2004)
Arsenic (As)	Mining, anthropogenic inputs	Disrupting soil structure and soil aggregates	Shrivastava *et al.* (2015)
Nickel (Ni)	Heavy metal, coal, diesel, oil, and anthropogenic contamination	Decreases soil pH	Iyaka (2011)
Cadmium (Cd)	Mining, smelting, industrial, vehicular emissions, manufacturing, and agrochemicals	Alter in soil cation exchange capacity and soil structure	Khan *et al.* (2017)
Mercury (Hg)	Coal combustion, waste incineration, and metal refining	Alter soil structure and water infiltration	Wang *et al.* (2012)
		Hydrocarbons	
Benzene	Pesticides, herbicides	Harms soil microorganism	Ezquerro *et al.* (2004)
Pesticides	Agricultural practices	Reduces soil aggregate formation and weakens the soil	Kodešová *et al.* (2010)
Vinyl chloride	Cleaners and degreasers	Adds toxic gases to soil	Eklund *et al.* (2022)

Table 1. (*Continued*)

Hazardous waste	Waste source	Effect on soil health	References
		Inorganic chemicals	
Hydrofluoric acid	Chemical plants	Loss of soil organic matter and nitrogen	An *et al.* (2014) and Sanderman *et al.* (2017)
Sulfuric acid	Chemical storage tank	Minerals like gypsum and kornelite, aluminite, and chloritoid forms	Sivapullaiah *et al.* (2009)

refineries, and ash from power plants that burn coal or incinerators. Waste from nuclear power reactors and nuclear weapons sites is the most toxic (Shukla *et al.*, 2020). When these types of wastes are dumped on the ground or in landfills, there is always a chance that groundwater will become contaminated because leachate produced by the disposal of untreated hazardous wastes is transported (Hasan *et al.*, 2021). Chemicals found in waste deposit leachates have the potential to alter the cohesive layer properties of soil, particularly pure clays. This is especially relevant to the long-term performance of mineral-based sealings (soil liners with low permeability). Chemicals that are classified as hazardous wastes include bleach, gasoline, batteries, and solvents, which are even found in household garbage and dying colors from the wastes of textile and food industries (Markandeya *et al.*, 2022; Garg and Chopra, 2021). Naturally occurring materials, such as heavy metals like lead and mercury, exposed to amounts far greater than usual due to human activity, can also be considered hazardous waste (Slonecker *et al.*, 2010).

The spread of harmful substances, acidification, salinization, and sodification, increased soil erosion with this chemical fertilizer, pesticides, fungicides, and the buildup of heavy metals and other inorganic pollutants are all factors that contribute to soil pollution. These substances alter the soil's physical and chemical properties (Weldeslassie *et al.*, 2017). Organic wastes are increasingly being added to agricultural soils as a disposal method, to enhance the physical and chemical characteristics of the soil, and to boost soil production (Abu-Rumman, 2016). Numerous organic wastes alter the physical characteristics of soil, including bulk density, saturated hydraulic conductivity, and water-holding capacity at both field capacity and wilting point. Recycling organic waste in agriculture may assist in preserving soil fertility through its influence on physical, chemical, and biological qualities (Al-Esawi *et al.*, 2020). Unsorted municipal household garbage intake in excess may cause changes in the physical and chemical properties of the soil. This may skew the interactions between the soil's chemical and biophysical functions (Anikwe and Nwobodo, 2002). Animal waste material concentrations have gradually raised several critical soil nutrients and contaminated groundwater (Akinbile *et al.*, 2016).

According to an Environmental Protection Agency (EPA) investigation, over 40,000 locations in the United States were determined to be contaminated with hazardous chemicals. In Northern France's grasslands, the density of earthworms is negatively impacted by heavy metal-enriched soil (Markandeya *et al.*, 2022). Around 70% of Denmark's municipal sewage sludge production in 1995 went to farming. This percentage has decreased since more people know that trace contaminants such as plasticizers and detergents can negatively affect soil and crop quality. Therefore, municipal sewage sludge production in Denmark dropped from about 170,000 tons per month in 1994 to 140,000 tons per month in 2002 (Jensen and Jepsen, 2004; Al-Esawi *et al.*, 2020). Whereas over the past three decades, China has faced significant hurdles in preserving its soil from contamination brought on by growing urbanization and industrialization. According to China's soil environmental quality limits, 16% of soil samples and 19% of agricultural soils are contaminated, primarily with heavy metals and metalloids, according to recent statewide studies. Nonetheless, it seems that the levels of certain heavy metals in Chinese soils are rising at far faster rates (Zhao *et al.*, 2014).

Since wastes are typically found in combinations, the conditions are varied and complex, with a gradient of contamination ranging from hot spots at e-waste sites to agricultural soils and mostly in open burning regions. Heavy metals and organic compounds (primarily polycyclic aromatic hydrocarbons (PAHs), polybrominated diphenyl ethers (PBDEs), polychlorinated and polybrominated biphenyls (PCBs and PBBs), dechlorane plus (DP), hexabromocyclododecanes (HBCDs), polychlorinated and polybrominated dibenzo-p-dioxins (PCDDs and PBDDs), and polychlorinated and polybrominated dibenzofurans (PCDFs and PBDFs)) are frequently found in e-waste areas (Zhang *et al.*, 2012). Waste transformer oil (WTO) is produced when the transformer's mechanical and electrical stresses cause its qualities to deteriorate while it is in use (Tiwari *et al.*, 2022) and contaminates soil. However, various anthropogenic activities in recent years have peeled away, sterilized, and poisoned the earth's soil with harmful chemicals (Katnoria and Sharma, 2019). Continuous municipal garbage disposal in soil may raise the concentration of heavy metals (Anikwe and Nwobodo, 2002). The critical relevance of food production and security makes the issue of high concentrations of HWs such as heavy metals, particularly in agricultural soils, a global environmental concern (Kowalska *et al.*, 2018). The origin of heavy metals may be inferred from their distribution within the soil profile. The presence of heavy metals in soil may reflect past human activity. However, the current sources of anthropogenic pollution, including industry, transportation, and agriculture, undoubtedly contribute to the buildup of heavy metals in the soil (Kowalska *et al.*, 2018). Since heavy metals and other hazardous chemicals can originate from nearby and far-off emission sources, they can either be deposited in their original location or travel great distances because of their capacity to bind to dust (Kowalska *et al.*, 2018). When such substances are present in the soil environment, the ecosystem becomes toxic and unbalanced (Katnoria and Sharma, 2019). Pollution indices are generally regarded as a helpful instrument for thoroughly assessing the level of contamination. Additionally, they can be very significant in determining the soil quality and forecasting the sustainability of future ecosystems, particularly regarding farmlands (Kowalska *et al.*, 2018). Control procedures must be implemented heroically to maintain the soil's fertility and productivity, which will benefit all living things (Aqeel *et al.*, 2014). Reuse, recycling, and waste minimization should be prioritized over disposal in any hazardous waste management plan. Low-cost hazardous waste management systems are particularly needed in developing nations (Orloff and Falk, 2003). This chapter will give a general review of contaminated soils, the physical and chemical changes caused by different hazardous wastes, remediation techniques, and a brief discussion of the significance of contaminated soil monitoring and evaluation.

2. Sources of Hazardous Waste in Soil

A substance has the ability to burn or ignite a fire if it is ignitable. Corrosive: Since it is so basic or acidic, it can dissolve steel or harm living creatures; reactive: it can detonate or emit harmful gases, such as sulfur or cyanide; it is categorized as a hazardous waste because it contains compounds that are poisonous to humans and other living things. Also, waste that is solid, liquid, sharp, or pathological is considered hazardous (Slonecker *et al.*, 2010; Samant *et al.*, 2018). Although hazardous waste makes up a small percentage of all garbage, it is the type that most threatens human health and the environment. Hazardous waste has a significant influence on the overall waste classification (Liang, 2023).

Hazardous wastes originate from particular locations or activities and are combined with soil, air, and water through everyday human activities or natural occurrences. Primary sources, secondary sources, point sources, and non-point sources are the four basic types of hazardous waste sources that pollute soil. According to primary sources, a variety of hazardous wastes are produced by the mining industry, construction, agriculture, industrial sectors, and medical equipment, and these pollutants are mixed with the soil during the deposition process. Hazardous material dependence is a reality in industrialized countries (Erkut *et al.*, 2006). Paper mill sludges, which are the most prevalent, and decarbonization sludges, which are generated by cleaning surface water, are two types of industrial wastes that are given particular consideration due to their significant differences from the other types of industrial sludge (Elskens *et al.*, 2013). Numerous activities, such as farming and animal husbandry, contribute to soil contamination. Soil pollution is a result of certain agricultural practices. These include fertilizers, certain agricultural methods, the use of long-lived insecticides, herbicides, fungicides, nematocides, etc., and animal feces included in agricultural waste (Aqeel *et al.*, 2014). Unsorted municipal household garbage intake in excess may cause changes in the physical and chemical properties of the soil. This may skew the interactions between the chemical and biophysical functions of soil (Anikwe and Nwobodo, 2002). Developed nations typically produce municipal solid wastes (MSWs) of 521.95 to 759.2 kg per person per year (kpc), while underdeveloped nations typically produce 109.5 to 525.6 (kpc). According to recent estimates, the world generates more than 2 billion tons of MSW annually, which could endanger environmental degradation (Karak *et al.*, 2012). Chemical, pesticide, and pharmaceutical manufacturing facilities frequently produce dangerous byproducts that, if improperly handled, might seep into the ground. Heavy metals, including lead, arsenic, and mercury, are released into the surrounding soil as a result of mining operations. Toxic chemicals can also be introduced by acid mine drainage. Heavy metals, solvents, and oily sludges are among the hazardous wastes produced by oil refineries

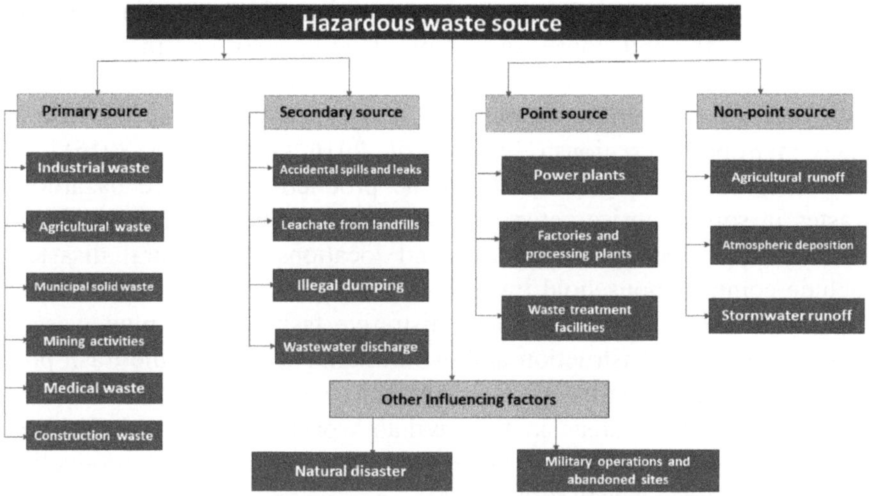

Fig. 1. Hazardous waste source classification.

that can pollute land in the event of spills or inappropriate trash disposal (Vambol *et al.*, 2019).

Secondary sources state that it includes leachate from landfills, some unlawful dumping, wastewater discharge to rivers or other natural waterbodies, and certain unintentional spills and leaks of oil or other hazardous chemicals in workplaces or during transportation (Fig. 1). Yet, unregulated and unlawful "landfills," also referred to as open dumpsites, are regrettably common in many developing nations. Since landfilling is still used so extensively, it is essential to look into any new health and environmental problems (Siddiqua *et al.*, 2022). Point and non-point sources are the other two varieties. Ash and slags produced by power plants from combustion wastes are known as point sources, and they contain arsenic and heavy metals. Hazardous chemical wastes, such as heavy metals, solvents, sulfur dioxide (SO_2), volatile organic compounds (VOCs), nitrogen oxides (NOx), and other harmful gases, can be produced by factories and processing facilities. Certain waste treatment facilities, such as landfills, treatment plants, and incineration, emit ammonia, hydrogen sulfide, ash, and percolated waste. These lead to a rise in soil degradation and waste from agriculture and industry. Toxic compounds found in the majority of them are bad for the air, water, land, and people (Chaaban, 2001). Non-point sources include stormwater runoff, atmospheric deposition, and agricultural runoff. These sources transport pollutants such as heavy metals, chemical particles, and excess nutrients from agricultural fields, increasing soil salinity. Atmospheric deposition adds acid rain or acidic pollutants to the soil, lowering soil pH. Due to these and the increased number and distance of vehicles on the road, concentrations of copper, zinc, and platinum group elements (PGEs) in sediments may increase in many metropolitan regions (Hwang *et al.*, 2016; Ghane *et al.*, 2016).

Other contributing elements to the production of more hazardous wastes in soil comprise natural disasters and military operations. Some military operations involving deserted locations and natural disasters include common household trash from residential regions, office wastes from different command and administrative tasks, landscaping wastes, and wastes from construction and demolition. As a result, the trash produced at military installations is comparable to that produced at commercial and residential areas in the civilian world (Medina and Waisner, 2011). For agriculture to provide enough food for the world, productive soils are required (Aqeel *et al.*, 2014).

3. Transport Mechanisms of Hazardous Wastes/ Contaminants in Soil

Numerous nations produce large quantities of various wastes (Table 2). Numerous causes, including metals, chemicals (monomers, solvents, reactive initiators, and dyes/pigments), leaching from chemical and municipal industries, unintentional spills, and subsurface leaks that pollute soil, can cause variations in the concentration of the heterogeneous substances. Industrial and commercial activities continuously generate HWs from a range of sources and processes. These pollutants can travel through the air and water and get mixed with soil (Markandeya *et al.*, 2022; Blagojevic *et al.*, 2022). HW transportation accidents are common, and this causes pollution of the environment or results in major injuries in populated areas (Batarliene, 2019). In the transportation of HWs from mining, petroleum refining, and chemical manufacturing in industry, various transport mechanisms such as diffusion, convection, advection, dispersion, dissolution, and adsorption play crucial roles. A number of variables, including temperature, fluid velocity, pH, sediment quantity, porosity, and others, influence these phenomena (Blagojevic *et al.*, 2022). HWs are not always found where they first appeared. HWs typically start in one location and move to other locations (Fig. 2). Hazardous material transportation can be divided into several categories based on the mode of

Table 2. Various countries highest garbage generation.

Country	Type of waste	Amount	Recycled/ biodegradable	References
India	Mixed waste	62 million tons	–	Maji *et al.* (2020)
Globally by 2019	E-waste	53.6 million tons	<13% recycled	Andeobu *et al.* (2021)
USA	Municipal solid wastes	(200–400) million tons/y	–	Tonjes and Greene (2012)
	Glass waste	75%	21% recycled	Ferdous *et al.* (2021)
	Tyre waste	75%	10% recycled	Ferdous *et al.* (2021)
	Plastic waste	55%	19.5% recycled	Ferdous *et al.* (2021)
Nigeria	Solid waste generation	0.129 kg/ capita/day	–	Afon and Okewole (2007)
Karachi, Pakistan	Municipal solid waste	12,000 tons/ day	71% biodegradable	Jilani and Rashid (2020)

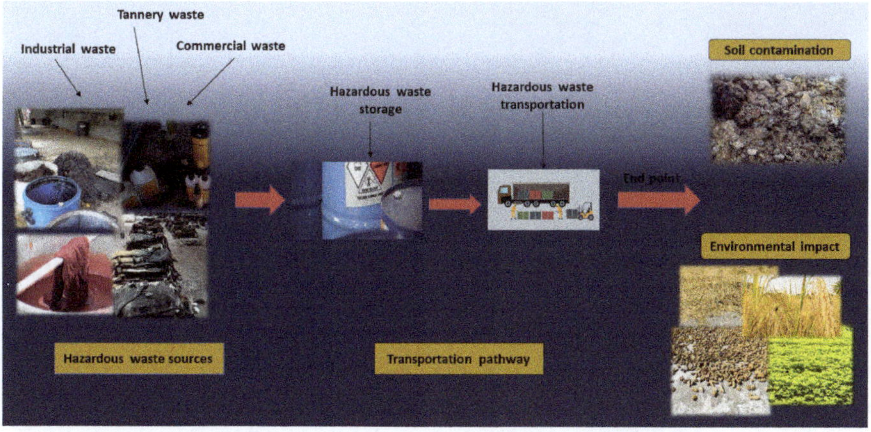

Fig. 2. Hazardous waste transportation pathway.

transportation, including pipeline, water, air, rail, and road. Intermodal shipments are those that are changed between modes while en route (Erkut *et al.*, 2006). For instance, oil is taken out of oil fields and transported to a refinery (usually by pipeline); numerous oil products, including gasoline and heating oil, are refined there before being transported to storage tanks spread out across a nation. Polychlorinated biphenyls (PCBs), for instance, are gathered from several industrial facilities, including former power generation and transfer stations, and transported to a specialized waste management facility for secure disposal (often cremation) (Erkut *et al.*, 2006).

A piece of land is considered contaminated if the amount of chemicals and other materials it contains exceeds a certain threshold, and the soil becomes dangerous for both crops and soil health. The main causes of contamination of soil are the deposition and interaction of various waste materials (solid, sludge, and liquid forms) (Nguyen *et al.*, 2020). Sewage sludge from urban wastewater treatment facilities and agro-industrial wastewater treatment sludge are the two primary types of sludge that are utilized as fertilizers. Since the pollutant concentrations of these sludge groups vary, they need to be investigated separately (Elskens *et al.*, 2013).

Toxic compounds, including polycyclic aromatic hydrocarbons (PAHs), polychlorinated biphenyls (PCBs), and heavy metals found in waste transformer oil (WTO), mingle with soil when it is disposed of in landfills, increasing the number of hydrocarbons and CO_2 in the soil and degrading its quality. Heavy metals in WTO alter the makeup and activity of soil aggregation and structure (Tiwari *et al.*, 2022).

4. Impact on Soil Physical Properties

Both agricultural output and the sustainable use of soil depend heavily on the physical properties (Table 3) of the soil. The capacity of the roots to both absorb and feed the soil solution to the roots determines the quantity and rate at which plants absorb water, oxygen, and nutrients. Low hydraulic conductivity, for example, can restrict the root's free flow of oxygen and water, which will harm agricultural productivity. De-oiled two-phase olive mill waste as a raw organic material might be the best amendment for enhancing the mechanical and physical characteristics of soil, which would increase the workable range and bring production back to deteriorated soils (López-Piñeiro *et al.*, 2011). Municipal solid waste in Nigeria affects the

Table 3. Wastes vs physical properties.

Different Wastes	**Porosity**	**Bulk density**	**Soil crusting**	**Water holding capacity**	**Water infiltration**	**Percolation**	**References**
Organic wastes	Increase	Decrease	Decrease	Increase	Increase	Increase	Almendro-Candel *et al.* (2018), Ozores-Hampton *et al.* (2011), and Abu-Rumman (2016)
Incorporated organic waste	No change	Decrease	No change	Increase	No change	No chance	Rezig *et al.* (2012)
Waste lubricating oil	Decrease	Increase	–	Decrease	–	–	Al-Esawi *et al.* (2020)
Olive mill wastes	–	Decrease	–	–	–	–	Salem and Ali (2023)
Municipal solid waste	Increase	Decrease	–	Increase	–	–	Zamani *et al.* (2015) and Anikwe and Nwobodo (2002)
Food waste	–	Increase	–	–	–	–	Reynolds *et al.* (2015)
Aquatic waste	–	Decrease	–	–	–	–	Jain and Kalamdhad (2020)
Animal waste	–	Decrease	–	–	–	–	Busari *et al.* (2012)
Pesticides	Increase	Decrease	–	–	–	–	Rajak *et al.* (2023)

bulk density of the soil, which is reduced by 9–13%, but the total porosity and hydraulic conductivity are increased by 9–14% and 240–463%, respectively, at the trash site (Anikwe and Nwobodo, 2002). Neither sewage sludge nor its compost has any negative effects on a number of soil properties that are consistent with higher soil fertility ratings. The physical characteristics of the soils are considerably enhanced by adding olive mill solid waste compared to the untreated soil (control). The total water intake and penetration depth increase by 30% and 29%, respectively, for clay soil and a 9% application rate. In sandy clay soil, the comparable increases are 32% and 25%. For clay and sandy clay soils, the soil (WHC) increases by 10.3% and 16.5%, respectively, at an application rate of 9%. The diluting effect of combining organic matter with denser soil minerals causes bulk density to decrease, but accumulation intake rises as the application rate increases (Abu-Rumman, 2016).

4.1. *Color, texture*

HWs contaminate soil by absorbing waste products, such as fertilizers, pesticides, sewage sludge, animal manures, and petroleum distillate spills (Darban *et al.*, 2022). Soil color and texture vary widely and are influenced by redox conditions, water content, and organic matter. Many of these physical characteristics are changed in soil when pesticide contamination is excessive (Rajak *et al.*, 2023). Heavy metal concentrations in agricultural areas are high due to the utilization of this untreated waste, which impacts the entire biosphere. They endanger both the plant and the food chain that eats it because plants immediately absorb them. They contaminate water and alter the soil's color, pH, and porosity. Consequently, removing hazardous heavy metals from contaminated wastewater is essential (Darban *et al.*, 2022).

4.2. *Structure, porosity, density, consistency, aggregate stability, and temperature*

4.2.1. *Structure*

One of the most significant physical characteristics of soil is its structure (Das *et al.*, 2011), which regulates or controls the movement and retention of water, solutes, gases, and biota in both natural and agricultural environments. An excessively acidic soil can reduce the availability of vital nutrients and increase the soil's susceptibility to erosion and deterioration of its structure (Rajak *et al.*, 2023). The arrangement of solid particles (texture) and empty spaces in a granular medium is referred to as its "structure." The majority of soils tend to be hierarchical. In other words, primary mineral particles typically form tiny clusters, or "first-order aggregates," in conjunction with organic molecules. These create "second-order aggregates," which are larger clusters. With each subsequent level, the aggregate size increases, reflecting the soil's aggregate hierarchy. Loams and clays are examples of medium- and fine-textured soils that would be almost impermeable to gases and liquids without a hierarchical structure. Particle rearrangement, flocculation, and cementation cause aggregation, a sign of soil structure (Almendro-Candel *et al.*, 2018).

4.2.2. *Porosity*

One important measure of the structural soundness of soil is porosity. A soil's pore space, or the area of its volume that is not occupied by solid material, is determined by its porosity and pore size distribution. Nearly everything that happens in the soil is governed by the fundamental characteristics of the pore space, including the flow of water, air, and other fluids; the transportation and reaction of chemicals; and the residency of roots and other biotas. Pesticides enhance soil porosity (Rajak *et al.*, 2023).

4.2.3. *Bulk density*

Soil bulk density, also known as dry bulk density, is one of the most well-known markers of soil structure. Porosity, compaction, productivity, and soil quality are all significantly influenced by bulk density. By increasing soil porosity, pesticides reduce the bulk density of the soil, which can exacerbate soil erosion and deterioration (Rajak *et al.*, 2023). It may be determined by sampling undisturbed soil and doesn't require any specialized knowledge or costly equipment. The dry mass of solids divided by the volume of soil is the bulk density. To determine soil porosity, bulk and particle density values are required. Once the particle density value is known or approximated, porosity can be calculated from bulk density (Almendro-Candel *et al.*, 2018). The impact of adding organic matter to the bulk density of soil shows that the bulk density decreases. This decrease in bulk density is explained by the diluting effect that occurs when the denser soil minerals are mixed with the extra organic matter (Abu-Rumman, 2016).

4.2.4. *Consistency*

The decreased consistency constraints of soil contaminated by industrial solid waste may be due to the decrease in surface area, which results in less water absorption (Khodary *et al.*, 2020). Oil waste can cover soil particles, but petroleum waste has a lubricating action and cannot alter the particle structure of the soil. The low consistency of the soil surface dominates the texture, the color turning dark black, and the pH. (Andriyan *et al.*, 2024).

4.2.5. *Aggregate stability*

One of the most important aspects of soil physical fertility is aggregate stability, which can be improved by managing organic amendments properly to preserve the proper soil structure. In contrast to soils with lower organic matter concentration, those with higher organic matter tend to contain larger, stronger, and more stable aggregates resistant to compaction. The primary factor influencing aggregate stability at the soil surface is exposure to rainwater (runoff and drop impact). When rains come into direct contact with bare soil that has had crop residues removed or mixed in by plowing, it breaks down soil aggregates more easily, making the soil more erodible. Surface sealing and crust development brought on by aggregate degradation can lower the water infiltration rate, raise the possibility of soil erosion, and result in the loss of important topsoil. For an agro ecosystem, increased soil aggregate stability has several benefits, such as a decreased chance of soil erosion and compaction (Almendro-Candel *et al.*, 2018)

4.2.6. *Temperature*

Direct disposal of hazardous wastes can change the moisture content and porosity of the soil because the heat produced by the chemical reaction or HWs at the dumpsite raises the soil's temperature (Yeşiller *et al.*, 2005).

5. Impact on Chemical Properties

Although some chemical elements are naturally present in soils as minerals, they can be harmful in certain amounts. Human activity on the soil may contain additional potentially dangerous materials (Vicent *et al.*, 2013).

The pulp and paper mill's effluents have a considerable impact on the soil mineralogy, as seen by the decline in silica levels, feldspar, and quartz contents, as well as the increase in iron and calcium contents at the sites (Adhikari and Bhattacharyya, 2015). Also, petroleum waste can contaminate soil and alter its chemical (Table 4) and physical characteristics, harming the soil ecology (Andriyan *et al.*, 2024). Compared to soil from non-dump sites, municipal solid waste changes the amount of soil organic matter, total nitrogen, cation exchange capacity, and percent base saturation rose from 701 to 743, 646 to 740, 7% to 10%, and 5% to 14%, respectively (Anikwe and Nwobodo, 2002; Abu-Rumman, 2016). It has been observed that arsenic (As) from compost rapidly dissociates after being surface-deposited as a mulch. Mobility through upper soil horizons to deeper, higher-pH technogenic layers was improved during periods of heavy rainfall. Concentrations of copper (Cu) in the soil underneath compost were much higher and strongly correlated with the amount of DOC that compost released into the pore water. Compost-derived DOC flushes also improved the mobility of soil's remaining cadmium (Cd) (Beesley and Dickinson, 2009).

5.1. *Mineralogy, organic matter content, cation exchange capacity, salinity, and pH*

5.1.1. *Mineralogy*

Since the salts in agricultural areas are more soluble, the soil becomes more fruitful when the ferric ion is reduced to the ferrous ion (Rajak *et al.*, 2023). Reduced conversion of ferric to ferrous ions due to pesticide contamination harms soil quality. Metal ions in agricultural areas have ligating sites, allowing them to form pesticide complexes. For example, thiophanate methyl can interact quickly with copper and nickel metal ions, while carbofuran forms complexes with iron and manganese ions. Reducing the amount of these components available to soil can damage the ecosystem and the integrity of the land.

5.1.2. *Organic matter content*

The retention of pesticides in agricultural lands is influenced by a number of critical soil qualities, including permeability, surface charge characteristics, and the amount of organic matter and clay in the soil (Rajak *et al.*, 2023).

Table 4. Wastes vs chemical properties.

Different wastes	Soil pH	Phosphorous content	Pottasium	Organic matter content	Moisture content (%)	Salinity	Cation exchange capacity	References
Waste lubricating oil	Decrease	Decrease	Decrease	Increase	Increase	-	-	Al-Esawi *et al*. (2020)
Organic wastes	–	Increase	Increase	Increase	Increase	Increases	Increase	Ozores-Hampton *et al*. (2011) and Abu-Rumman (2016)
Olive mill wastes	Decrease	–	–	–	–	Decrease	–	Salem and Ali (2023)
Rice husk and tobacco waste	Increase	Increase	Increase	Increase	–	–	Increase	Hamzah *et al*. (2017)
Aquatic waste	Increase	–	–	–	–	–	–	Jain and Kalamdhad (2020)
Municipal solid waste	Increase	–	–	Increase	–	Increase	Increase	Jilani and Rashid (2020) and Anikwe and Nwobodo (2002)

Furthermore, using pesticides on agricultural fields has a detrimental effect on the organic contents of the soil and microbial biomass. Industrial solid waste slightly decreases the soil organic matter content (Khodary *et al.*, 2020).

5.1.3. *Cation exchange capacity*

Compared to other uncontaminated site soil, HW may improve the chemical characteristics of the dump site soil, such as electrical conductivity and cation exchange capacity (Jing *et al.*, 2019). Rice husk and tobacco trash may also enhance the cation exchange capacity and soil organic carbon. If soil organic matter and cation exchange capacity increased, cation availability would also improve (Hamzah *et al.*, 2017).

5.2. *Soil salinity and soil pH*

5.2.1. *Salinity*

Potential chemical benefits of urban waste components in restoring degraded soils and promoting plant growth may be outweighed by chemical issues, including elevated soil salinity and heavy metal buildup (Albaladejo *et al.*, 1994). Compared to the control treatment, the compost from olive mill waste treatment at a rate of 10 tons per hectare dramatically reduced pH, electric conductivity (EC), and soil bulk density by 18.7 and 20%, respectively (Salem and Ali, 2023).

5.2.2. *Soil pH*

Soil pH is an important soil parameter that affects pesticide retention in agricultural soil. Empty pesticide containers can result in soil–pesticide interactions (Rajak *et al.*, 2023). De Sousa Oliveira *et al.* (2018) found that, when compared to other chemical elements, shrimp aquaculture had the lowest organic carbon concentration. The compounds derived from small ruminant residues and poultry industry residues had the highest quantities of phosphate and nitrogen. Even though not all soils are alkaline, applying rice husk and tobacco waste improves the alkalinity and raises the pH of the soil (Hamzah *et al.*, 2017).

6. Monitoring and Assessment of Soil Contamination

Hazardous waste management is a major problem in our country right now. The dearth of safe landfill locations in the country for the environmentally responsible disposal of hazardous waste and the absence of scientific approaches for hazardous waste disposal pose a major threat to the environmental system. First, industrialization over the last few decades has led to a decrease in natural resources and an increase in pollution. From that time, waste disposal has been the root cause of many issues, such as inadequate treatment opportunities, ecological security, illegal waste dumping, a lack of paved or asphalt access roads to the landfill, and low efficiency of public waste management services; environmentalists now view waste disposal as being extremely complex (Vambol *et al.*, 2019). Landfilling is one of the most widely employed waste management strategies in all countries, regardless of their degree of development. The three most common landfill kinds are hazardous, industrial, and municipal solid waste dumps. Furthermore, a recently created landfill type called a "green waste landfill" is sometimes appropriate (Siddiqua *et al.*, 2022). Landfilling is the most often used method of garbage disposal since it is inexpensive. A liner, which stops fluid flow and protects the groundwater and underlying land, is used when disposing of any hazardous material in a landfill. Among the various methods for boosting and improving the characteristics of landfills, laboratory research has looked into methods that use bentonite, cement, lime, gypsum, and other materials (Saha *et al.*, 2008).

By implementing sustainable methods, hazardous waste contamination monitoring and assessment (HWCMA) can prevent the long-term health consequences of waste contamination on soil and preserve ecosystem health, as well as physical, chemical, and soil health. Six steps are involved (Fig. 3). The first stage is to identify the possible sources of hazardous waste contamination sites, such as landfills, industrial sites, and agricultural areas. After that, a site survey and inspections (including aerial photography, historical record reviews, and visual inspections) ought to be carried out. It is necessary to gather and describe biological samples as well as environmental samples, such as soil, water, and air. The potential risk should be evaluated in the second stage. Here, it is important to identify the possible exposure pathway for hazards. Ingestion, inhalation, and skin contact can all be examined, and possible health concerns or threats to human health can be evaluated, along with the overall effects

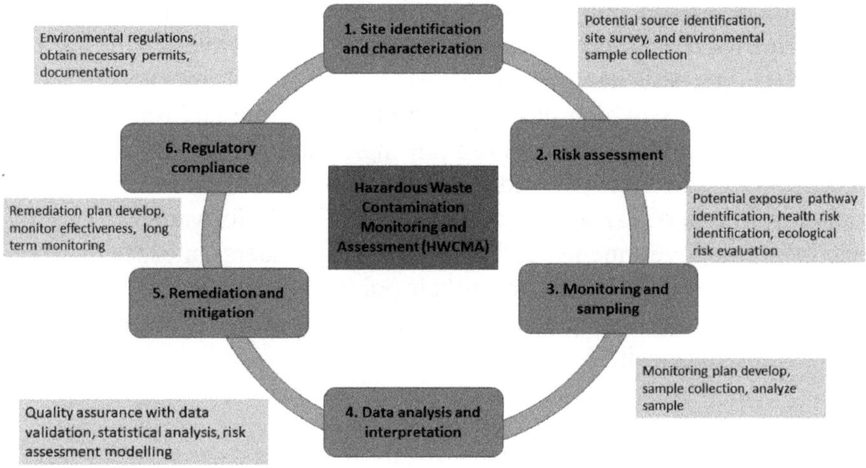

Fig. 3. Hazardous waste contamination monitoring and assessment (HWCMA).

on the environment. Step three should begin with the development of a monitoring plan that includes laboratory analysis to determine sample locations, frequency, and parameters. Errors and inconsistencies would then be found in the analysis of the collected data. The fourth phase could possibly involve some significant statistical analysis.

Modeling for risk assessment can also be incorporated. By putting cleanup plans into action and keeping an eye on their efficacy, remediation and mitigation can be accomplished while lowering the level of basic contaminants in the soil. Last but not least, regulatory compliance should be completed, encompassing federal, state, and local regulations along with the required paperwork and permissions. This will clarify further how the environmental regulation guarantees responsible management and reduction of HW from the risk of soil contamination. The prevention of hazardous wastes and the improvement of industrial efficiency depend on recycling. Additionally, it is required to oversee the disposal of massive amounts of hazardous waste (Chaaban, 2001). Eliminating hazardous waste involves methods and procedures that lessen the need for hazardous waste treatment either before it enters a receiving environment or when it is finally disposed of (Chaaban, 2001; Batarliene, 2019; Qian *et al.*, 2021). It is crucial to separate all types of HWs from the regular waste stream and to treat and transport them appropriately. Risk analysis can be conducted using a geographic information system (GIS). The federal government should put in place certain laws and regulations to lessen hazardous waste and its detrimental impact on the environment. If hazardous wastes are still handled, transported, stored, and disposed of in an unsystematic way, which has negative effects on the environment (soil, water, and air) and human health, some significant steps should be taken. Hazardous waste transportation should involve a number of partners, including shippers, carriers, package makers, freight forwarders, consignees, insurers, governments, and emergency responders, in order to safely transfer hazardous materials from their points of origin to their final destinations. They must keep a close eye out for any leaks or incorrect packaging while in transit.

7. Conclusion

A significant substitute for landfilling is the use of organic wastes as soil supplements, which most of the time improve soil structure, water retention, and the concentrations of organic matter and nutrients (Renaud *et al.*, 2017). But a lot of toxic chemicals or matters of waste alter the physical and chemical properties of soil. Controlling all these transportations of HWs and minimizing environmental pollution depend on efficient waste management. Risks associated with the transportation of hazardous materials can be decreased in a variety of ways. A few of these risk-reduction strategies, such as routine maintenance, evacuation, routine monitoring, driver education, and increasing the knowledge of the detrimental effects of HWs to workers or all the people who are involved with hazardous materials directly or indirectly, and operations research could be developed. To make people more conscious of their use, both domestically and internationally, local people and workers must be taught about dangerous substances and materials as well as the negative consequences of contaminated soil. Strict legislation must be passed in order to enforce compliance and reduce the amount of hazardous waste produced.

Bibliography

Abu-Rumman, G. (2016). Effect of olive mill solid waste on soil physical properties. *International Journal of Soil Science*, 11, 94–101. https://doi.org/10.3923/ijss.2016.94.101.

Adhikari, G. and Bhattacharyya, K.G. (2015). Impact of pulp and paper mill effluents and solid wastes on soil mineralogical and physicochemical properties. *Environmental Monitoring and Assessment*, 187(3). https://doi.org/10.1007/s10661-015-4330-z.

Afon, A.O. and Okewole, A. (2007). Estimating the quantity of solid waste generation in Oyo, Nigeria. *Waste Management & Research*, 25, 371–379. https://doi.org/10.1177/0734242x07078286.

Akinbile, C.O., Erazua, A.E., Babalola, T.E., and Ajibade, F.O. (2016). Environmental implications of animal wastes pollution on agricultural soil and water quality. *Soil and Water Research*, 11, 172–180. https://doi.org/10.17221/29/2015-swr.

Albaladejo, J., Stocking, M., Diaz, E., and Castillo, V. (1994). Land rehabilitation by urban refuse amendments in a semi-arid environment: Effect on soil chemical properties. *Soil Technology*, 7, 249–260. https://doi.org/10.1016/0933-3630(94)90025-6.

Al-Esawi, J.S.E., Wayayok, A., Al-Ogaidi, A.A.M., Rowshon, M.K., Abdullah, A.F., and Abdullahi, S. (2020). Effect of soil compaction and palm oil application on soil infiltration rate. *Journal of Irrigation and Drainage Engineering*, 147(3). https://doi.org/10.1061/(asce)ir.1943-4774.0001534.

Almendro-Candel, M.B., Lucas, I.G., Navarro-Pedreño, J., and Zorpas, A.A. (2018). Physical properties of soils affected by the use of agricultural waste. In: *Physical Properties of Soils Affected by the Use of Agricultural Waste*. IntechOpen. https://doi.org/10.5772/intechopen.77993.

An, J., Lee, H.A., Lee, J., and Yoon, H. (2014). Fluorine distribution in soil in the vicinity of an accidental spillage of hydrofluoric acid in Korea. *Chemosphere*, 119, 577–582. https://doi.org/10.1016/j.chemosphere.2014.07.043.

Andeobu, L., Wibowo, S., and Grandhi, S. (2021). An assessment of e-waste generation and environmental management of selected countries in Africa, Europe and North America: A systematic review. *Science of the Total Environment*, 792, 148078. https://doi.org/10.1016/j.scitotenv.2021.148078.

Andriyan, R., Astutik, S., Wijayanto, Y., Kurnianto, F.A., and Pangastuti, E.I. (2024). The analysis of the impact of petroleum mining on the environment and community economy in Wonocolo Village of Bojonegoro Regency. *Jurnal Kesehatan Lingkungan Indonesia*, 23, 279–286. https://doi.org/10.14710/jkli.23.3.280-287.

Anikwe, M. and Nwobodo, K. (2002). Long term effect of municipal waste disposal on soil properties and productivity of sites used for urban agriculture in Abakaliki, Nigeria. *Bioresource Technology*, 83, 241–250. https://doi.org/10.1016/s0960-8524(01)00154-7.

Aqeel, M., Jamil, M., and Yusoff, I. (2014). Soil contamination, risk assessment and remediation. In: *Soil Contamination, Risk Assessment and Remediation*. IntechOpen. https://doi.org/10.5772/57287.

Batarliene, N. (2019). Research on transportation of health-hazardous waste. *European Journal of Sustainable Development Research*, 4(1). https://doi.org/10.29333/ejosdr/6342.

Beesley, L. and Dickinson, N. (2009). Carbon and trace element mobility in an urban soil amended with green waste compost. *Journal of Soils and Sediments*, 10, 215–222. https://doi.org/10.1007/s11368-009-0112-y.

Blagojevic, M., Buchal, S., Chute, M., Shareefdeen, Z., and Elkamel, A. (2022). Hazardous waste transport in the environment. In: *Environmental Sustainability in the Oil and Gas Industry*. Springer, pp. 273–295. https://doi.org/10.1007/978-3-030-95262-4_11.

Bundhoo, Z.M.A. (2018). Solid waste management in least developed countries: current status and challenges faced. *Journal of Material Cycles and Waste Management*, 20, 1867–1877. https://doi.org/10.1007/s10163-018-0728-3.

Busari, M., Azeez, J., and Orelaja, O. (2012). Impact of long term (9 years) deposition of animal wastes on soil physical properties in Abeokuta, South-Western Nigeria—implications for soil management. *The Journal of Solid Waste Technology and Management*, 38, 271–278. https://doi.org/10.5276/jswtm.2012.271.

Chaaban, M.A. (2001). Hazardous waste source reduction in materials and processing technologies. *Journal of Materials Processing Technology*, 119, 336–343. https://doi.org/10.1016/s0924-0136(01)00920-7.

Darban, Z., Shahabuddin, S., Gaur, R., Ahmad, I., and Sridewi, N. (2022). Hydrogel-based adsorbent material for the effective removal of heavy metals from wastewater: A comprehensive review. *Gels*, 8(5), 263. https://doi.org/10.3390/gels8050263.

Das, A., Sherameti, I., and Varma, A. (2011). Contaminated soil: Physical, chemical and biological components. In: *Soil Biology* (pp. 1–15). https://doi.org/10.1007/978-3-642-23327-2_1.

De Sousa Oliveira, L., Costa, M.C.G., De Souza, H.A., Blum, J., Da Silva Albuquerque, G.H., De Abreu, M.G.P., and De Sousa Maia, D. (2018). Characterization of organic wastes and effects of their application on the soil. *Journal of Agricultural Science*, 10, 291. https://doi.org/10.5539/jas.v10n6p291.

Dixit, A. and Srivastava, R. (2015). An estimate of contaminated land area due to industrial hazardous waste generation in India. *International Journal of Advanced Research in Education & Technology*, 2(3), 117–125.

Eklund, B., Rago, R., Plantz, G., Haddad, E., Miesfeldt, M., and Volpi, R. (2022). Fate & transport of vinyl chloride in soil vapor. *Remediation Journal*, 32, 273–279. https://doi.org/10.1002/rem.21732.

Elskens, M., Pussemier, L., Dumortier, P., Van Langenhove, K., Scholl, G., Goeyens, L., and Focant, J. (2013). Dioxin levels in fertilizers from Belgium:

Determination and evaluation of the potential impact on soil contamination. *Science of the Total Environment*, 454–455, 366–372. https://doi.org/10.1016/j.scitotenv.2013.02.094.

Erkut, E., Tjandra, S.A., and Verter, V. (2006). Hazardous materials transportation. In: *Handbooks in Operations Research and Management Science* (pp. 539–621). https://doi.org/10.1016/s0927-0507(06)14009-8.

Ezquerro, O., Ortiz, G., Pons, B., and Tena, M.T. (2004). Determination of benzene, toluene, ethylbenzene and xylenes in soils by multiple headspace solid-phase microextraction. *Journal of chromatography A*, 1035(1), 17–22.

Ferdous, W., Manalo, A., Siddique, R., Mendis, P., Zhuge, Y., Wong, H.S., Lokuge, W., Aravinthan, T., and Schubel, P. (2021). Recycling of landfill wastes (tyres, plastics and glass) in construction—A review on global waste generation, performance, application and future opportunities. *Resources, Conservation and Recycling*, 173, 105745. https://doi.org/10.1016/j.resconrec.2021.105745.

Garg, A. and Chopra, L. (2021). Dye waste: A significant environmental hazard. *Materials Today: Proceedings*, 48, 1310–1315. https://doi.org/10.1016/j.matpr.2021.09.003.

Ghane, E., Ranaivoson, A.Z., Feyereisen, G.W., Rosen, C.J., and Moncrief, J.F. (2016). Comparison of contaminant transport in agricultural drainage water and urban stormwater runoff. *PLoS ONE*, 11(12), e0167834. https://doi.org/10.1371/journal.pone.0167834.

Hamzah, A., Hapsari, R.I., and Priyadarshini, R. (2017). The influence of rice husk and tobacco waste biochars on soil quality. *Journal of Degraded and Mining Lands Management*, 5, 1001–1007. https://doi.org/10.15243/jdmlm.2017.051.1001.

Hasan, M.A., Ahmad, S., and Mohammed, T. (2021). Groundwater contamination by hazardous wastes. *Arabian Journal for Science and Engineering*, 46, 4191–4212. https://doi.org/10.1007/s13369-021-05452-7.

Holtzer, M., Kmita, A., Żymankowska-Kumon, S., Bobrowski, A., and Dańko, R. (2016). Influence of the hardener on the emission of harmful substances from moulding sands with furan resin in the pyrolysis process. *Archives of Foundry Engineering*, 16, 107–111. https://doi.org/10.1515/afe-2016-0012.

Hwang, H., Fiala, M.J., Park, D., and Wade, T.L. (2016). Review of pollutants in urban road dust and stormwater runoff: part 1. Heavy metals released from vehicles. *International Journal of Urban Sciences*, 20, 334–360. https://doi.org/10.1080/12265934.2016.1193041.

Iyaka, N.Y.A. (2011). Nickel in soils: A review of its distribution and impacts. *Scientific Research and Essays*, 6(33). https://doi.org/10.5897/srex11.035.

Jain, M.S. and Kalamdhad, A.S. (2020). Soil revitalization via waste utilization: Compost effects on soil organic properties, nutritional, sorption and physical properties. *Environmental Technology & Innovation*, 18, 100668. https://doi.org/10.1016/j.eti.2020.100668.

Jensen, J. and Jepsen, S. (2004). The production, use and quality of sewage sludge in Denmark. *Waste Management*, 25, 239–247. https://doi.org/10.1016/j.wasman.2004.08.011.

Jilani, S. and Rashid, R. (2020). Municipal solid waste dumping and its impact on soil quality in Karachi. *DOAJ (DOAJ: Directory of Open Access Journals)*. https://doi.org/10.6092/issn.2281-4485/9342.

Jing, F., Yang, Z., Chen, X., Liu, W., Guo, B., Lin, G., Huang, R., and Liu, W. (2019). Potentially hazardous element accumulation in rice tissues and their availability in soil systems after biochar amendments. *Journal of Soils and Sediments*, 19, 2957–2970. https://doi.org/10.1007/s11368-019-02296-5.

Karak, T., Bhagat, R.M., and Bhattacharyya, P. (2012). Municipal solid waste generation, composition, and management: The world scenario. *Critical Reviews in Environmental Science and Technology*, 42, 1509–1630. https://doi.org/10.1080/10643389.2011.569871.

Katnoria, J.K. and Sharma, P. (2019). Impacts of hazardous waste on soil health. In: *Practice, Progress, and Proficiency in Sustainability* (pp. 18–35). https://doi.org/10.4018/978-1-7998-0031-6.ch002.

Khaleel, R., Reddy, K.R., and Overcash, M.R. (1981). Changes in soil physical properties due to organic waste applications: A review. *Journal of Environmental Quality*, 10, 133–141. https://doi.org/10.2134/jeq1981.00472425001000020002x.

Khan, M.A., Khan, S., Khan, A., and Alam, M. (2017). Soil contamination with cadmium, consequences and remediation using organic amendments. *Science of the Total Environment*, 601–602, 1591–1605. https://doi.org/10.1016/j.scitotenv.2017.06.030.

Khodary, S.M., Elwakil, A.Z., Fujii, M., and Tawfik, A. (2020). Effect of hazardous industrial solid waste landfill leachate on the geotechnical properties of clay. *Arabian Journal of Geosciences*, 13(15). https://doi.org/10.1007/s12517-020-05699-8.

Kodešová, R., Kočárek, M., Kodeš, V., Drábek, O., Kozák, J., and Hejtmánková, K. (2010). Pesticide adsorption in relation to soil properties and soil type distribution in regional scale. *Journal of Hazardous Materials*, 186, 540–550. https://doi.org/10.1016/j.jhazmat.2010.11.040.

Kowalska, J.B., Mazurek, R., Gąsiorek, M., and Zaleski, T. (2018). Pollution indices as useful tools for the comprehensive evaluation of the degree of soil contamination—A review. *Environmental Geochemistry and Health*, 40, 2395–2420. https://doi.org/10.1007/s10653-018-0106-z.

Liang, J. (2023). Characteristics of hazardous waste and recovery. *Highlights in Science, Engineering and Technology*, 59, 91–96. https://doi.org/10.54097/hset.v59i.10065.

López-Piñeiro, A., Albarrán, A., Nunes, J.R., Peña, D., and Cabrera, D. (2011). Long-term impacts of de-oiled two-phase olive mill waste on soil chemical properties, enzyme activities and productivity in an olive grove. *Soil and Tillage Research*, 114, 175–182. https://doi.org/10.1016/j.still.2011.05.002.

Maji, S., Dwivedi, D.H., Singh, N., Kishor, S., and Gond, M. (2020). Agricultural waste: Its impact on environment and management approaches. In: *Microorganisms for Sustainability* (pp. 329–351). https://doi.org/10.1007/978-981-15-1390-9_15.

Markandeya, N., Mohan, D., and Shukla, S.P. (2022). Hazardous consequences of textile mill effluents on soil and their remediation approaches. *Cleaner Engineering and Technology*, 7, 100434. https://doi.org/10.1016/j.clet.2022.100434.

Medina, V.F. and Waisner, S.A. (2011). Military solid and hazardous wastes—Assessment of issues at military facilities and base camps. In: *Elsevier eBooks* (pp. 357–376). https://doi.org/10.1016/b978-0-12-381475-3.10025-7.

Nguyen, H.T., Phuong, V.N., Van, T.N., Thi, P.N., Lan, P.D.T., Pham, H.T., and Cao, H.T. (2020). Low-cost hydrogel derived from agro-waste for veterinary antibiotic removal: Optimization, kinetics, and toxicity evaluation. *Environmental Technology & Innovation*, 20, 101098. https://doi.org/10.1016/j.eti.2020.101098.

Orloff, K. and Falk, H. (2003). An international perspective on hazardous waste practices. *International Journal of Hygiene and Environmental Health*, 206, 291–302. https://doi.org/10.1078/1438-4639-00225.

Ozores-Hampton, M., Stansly, P.A., and Salame, T.P. (2011). Soil chemical, physical, and biological properties of a sandy soil subjected to long-term organic amendments. *Journal of Sustainable Agriculture*, 35, 243–259. https://doi.org/10.1080/10440046.2011.554289.

Qian, G., Duanmu, C., Ali, N., Khan, A., Malik, S., Yang, Y., and Bilal, M. (2021). Hazardous wastes, adverse impacts, and management strategies: A way forward to environmental sustainability. *Environment, Development and Sustainability*, 24, 9731–9756. https://doi.org/10.1007/s10668-021-01867-2.

Rajak, P., Roy, S., Ganguly, A., Mandi, M., Dutta, A., Das, K., Nanda, S., Ghanty, S., and Biswas, G. (2023). Agricultural pesticides—friends or foes to biosphere? *Journal of Hazardous Materials Advances*, 10, 100264. https://doi.org/10.1016/j.hazadv.2023.100264.

Renaud, M., Chelinho, S., Alvarenga, P., Mourinha, C., Palma, P., Sousa, J.P., and Natal-Da-Luz, T. (2017). Organic wastes as soil amendments—Effects assessment towards soil invertebrates. *Journal of Hazardous Materials*, 330, 149–156. https://doi.org/10.1016/j.jhazmat.2017.01.052.

Reynolds, W.D., Drury, C.F., Tan, C.S., and Yang, X.M. (2015). Temporal effects of food waste compost on soil physical quality and productivity. *Canadian Journal of Soil Science*, 95, 251–268. https://doi.org/10.4141/cjss-2014-114.

Rezig, A.M.R., Elhadi, E.A., and Mubarak, A.R. (2012). Effect of incorporation of some wastes on a wheat-guar rotation system on soil physical and chemical properties. *International Journal of Recycling of Organic Waste in Agriculture*, 1(1). https://doi.org/10.1186/2251-7715-1-1.

Saha, P., Datta, S., and Sanyal, S.K. (2008). Hazardous waste pollution prevention using clay with admixtures. *Clean—Soil, Air, Water*, 36, 230–238. https://doi.org/10.1002/clen.200700086.

Salem, H.M. and Ali, A.M. (2023). Effect of olive mill wastes on soil physicochemical properties and maize yield under saline soil conditions. *Journal of Soil, Plant and Environment*, 2, 53–64. https://doi.org/10.56946/jspae.v2i1.156.

Samant, M., Pandey, S.C., and Pandey, A. (2018). Impact of hazardous waste material on environment and their management strategies. In: *Advances in Environmental Engineering and Green Technologies* (pp. 175–192). https://doi.org/10.4018/978-1-5225-3126-5.ch011.

Sanderman, J., Farrell, M., Macreadie, P.I., Hayes, M., McGowan, J., and Baldock, J. (2017). Is demineralization with dilute hydrofluoric acid a viable method for isolating mineral stabilized soil organic matter? *Geoderma*, 304, 4–11. https://doi.org/10.1016/j.geoderma.2017.03.002.

Shrivastava, A., Ghosh, D., Dash, A., and Bose, S. (2015). Arsenic contamination in soil and sediment in India: Sources, effects, and remediation. *Current Pollution Reports*, 1, 35–46. https://doi.org/10.1007/s40726-015-0004-2.

Shukla, S., Ganguly, R., and Hussain, C.M. (2020). Hazardous wastes—types and sources. In: *Royal Society of Chemistry eBooks* (pp. 24–52). https://doi.org/10.1039/9781788016261-00024.

Siddiqua, A., Hahladakis, J.N., and Al-Attiya, W.A.K.A. (2022). An overview of the environmental pollution and health effects associated with waste landfilling and open dumping. *Environmental Science and Pollution Research*, 29, 58514–58536. https://doi.org/10.1007/s11356-022-21578-z.

Sivapullaiah, P.V., Prasad, B.G., and Allam, M.M. (2009). Effect of sulfuric acid on swelling behavior of an expansive soil. *Soil and Sediment Contamination*, 18, 121–135. https://doi.org/10.1080/15320380802660289.

Slonecker, T., Fisher, G.B., Aiello, D.P., and Haack, B. (2010). Visible and infrared remote imaging of hazardous waste: A review. *Remote Sensing*, 2, 2474–2508. https://doi.org/10.3390/rs2112474.

Soares, A.A., Pinho, M.T., Albergaria, J.T., Domingues, V., Da Conceição M Alvim-Ferraz, M., De Marco, P., and Delerue-Matos, C. (2011). Sequential application of soil vapor extraction and bioremediation processes for the remediation of ethylbenzene-contaminated soils. *Water, Air, & Soil Pollution*, 223, 2601–2609. https://doi.org/10.1007/s11270-011-1051-y.

Testa, S. (2004). Sources of chromium contamination in soil and groundwater. In: *CRC Press eBooks* (pp. 143–163). https://doi.org/10.1201/9780203487969.ch4.

Tiwari, R., Agrawal, P., Bawa, S., Karadbhajne, V., and Agrawal, A.J. (2022). Soil contamination by waste transformer oil: A review. *Materials Today: Proceedings*, 72, 306–310. https://doi.org/10.1016/j.matpr.2022.07.403.

Tonjes, D.J. and Greene, K.L. (2012). A review of national municipal solid waste generation assessments in the USA. *Waste Management & Research*, 30, 758–771. https://doi.org/10.1177/0734242x12451305.

Vambol, S., Vambol, V., Sundararajan, M., and Ansari, I. (2019). The nature and detection of unauthorized waste dump sites using remote sensing. *Ecological Questions*, 30, 1. https://doi.org/10.12775/eq.2019.018.

Van Straalen, N.M. (2002). Assessment of soil contamination—a functional perspective. *Biodegradation*, 13, 41–52. https://doi.org/10.1023/a:1016398018140.

Vicent, T., Caminal, G., Eljarrat, E., and Barceló, D. (2013). Emerging organic contaminants in sludges. In: *The Handbook of Environmental Chemistry*. https://doi.org/10.1007/978-3-642-35609-4.

Wang, J., Feng, X., Anderson, C.W., Xing, Y., and Shang, L. (2012). Remediation of mercury contaminated sites—A review. *Journal of Hazardous Materials*, 221–222, 1–18. https://doi.org/10.1016/j.jhazmat.2012.04.035.

Weldeslassie, T., Naz, H., Singh, B., and Oves, M. (2017). Chemical contaminants for soil, air and aquatic ecosystem. In: *Springer eBooks* (pp. 1–22). https://doi.org/10.1007/978-3-319-64501-8_1.

Yeşiller, N., Hanson, J.L., and Liu, W. (2005). Heat generation in municipal solid waste landfills. *Journal of Geotechnical and Geoenvironmental Engineering*, 131, 1330–1344. https://doi.org/10.1061/(asce)1090-0241(2005)131:11(1330.

Zamani, J., Afyuni, M., Sepehrnia, N., and Schulin, R. (2015). Opposite effects of two organic wastes on the physical quality of an agricultural soil. *Archives of Agronomy and Soil Science*, 62, 413–427. https://doi.org/10.1080/03650340.2015.1060321.

Zeeb, B.A., Amphlett, J.S., Rutter, A., and Reimer, K.J. (2006). Potential for phytoremediation of polychlorinated biphenyl-(PCB)-contaminated soil. *International Journal of Phytoremediation*, 8, 199–221. https://doi.org/10.1080/15226510600846749.

Zhang, W., Wu, Y., and Simonnot, M. (2012). Soil contamination due to e-waste disposal and recycling activities: A review with special focus on China. *Pedosphere*, 22, 434–455. https://doi.org/10.1016/s1002-0160(12)60030-7.

Zhao, F., Ma, Y., Zhu, Y., Tang, Z., and McGrath, S.P. (2014). Soil contamination in China: Current status and mitigation strategies. *Environmental Science & Technology*, 49, 750–759. https://doi.org/10.1021/es5047099.

https://doi.org/10.1142/9789819822249_0004

Chapter 4

Health Risks Linked to Hazardous Waste-Related Metal Contamination of Food Crops

Samuel Oluwasegun Adesida[*,‡] **and Chibuisi Gideon Alimba**[†]

[*]*Department of Environmental Protection in Agriculture, Faculty of Agriculture, University of Belgrade, Belgrade, Serbia*

[†]*IU Internationale Hochschule, Buchdorf, Germany*

[‡]*samueladesida43@gmail.com*

Abstract

The management and disposal of hazardous wastes, a priority concern for Member States of the World Health Organization (WHO) Regional Office for Europe, are critical issues due to their profound impacts on public health and environmental sustainability. Hazardous waste is primarily characterized by its toxicity, with persistent heavy metals being one of its most harmful constituents. The generation of hazardous waste as a byproduct of industrial operations or abandoned commercial products has resulted in an abrupt rise in human exposure to metal contaminants. Human exposure to heavy metals through the consumption of contaminated food crops represents a significant public health challenge, as these metals are known to pose both carcinogenic and non-carcinogenic health risks. This chapter presents information on metal bioaccumulation in food crops planted within the vicinity of hazardous waste sites and the possible carcinogenic and non-carcinogenic health implications of consuming such metal-laden food crops. It focused on specific metals including, arsenic, lead, cadmium, chromium, and mercury, which are on the priority list of hazardous substances capable of causing permanent harmful effects on human and wildlife. An in-depth discussion of critical pathways and possible mechanisms of induction of health hazard is also presented. By exploring the intersection of environmental (metal) pollution from hazardous waste management, food safety/security, and public health, a broader understanding of the need for sustainable waste management practices and stringent regulatory frameworks to mitigate metal pollution in agricultural soils is established.

Keywords: Hazardous waste, metal pollution, soil contamination, metal bioaccumulation in food crops, metal bioconcentration in tertiary consumer, health consequences.

1. Introduction

The issue of hazardous waste management has emerged as a critical environmental and public health priority around the globe, particularly for Member States of the World Health Organization (WHO) Regional Office for Europe. This concern was notably addressed during the Sixth Ministerial Conference on Environment and Health (WHO, 2016), reflecting the urgency with which nations are dealing with the environmental and health risks posed by hazardous waste. The quantity of hazardous waste generated varies considerably across nations, often correlating with levels of industrialization. For instance, in 1998, the Russian Federation generated approximately 107 million metric tons of hazardous waste, while Uganda produced only 38.3 metric tons (Basel, 1998). While non-industrialized nations typically generate lower volumes of hazardous waste, the rapid industrialization of the globe has exacerbated the challenges associated with the management of this toxic byproduct. The U.S. Environmental Protection Agency (EPA) estimated that 10–15% of the total municipal and industrial solid wastes produced in the U.S. in 1981 were classified as hazardous (Polprasert, 1996). In China, between 1999 and 2005, the National Bureau of Statistics of China estimated that 70,000 tons of industrial hazardous wastes were released into the environment without proper control measures (Duan *et al.*, 2008). In January 2007, the

US Environmental Protection Agency's National Priority List (NPL) identified 1240 hazardous waste sites, including 157 federal establishments (Fazzo *et al.*, 2017). Such figures illustrate the global scale of hazardous waste production and the difficulties in its management.

Hazardous waste is generated from a wide array of sources, including industrial operations, small enterprises, households, hospitals, research laboratories, and agriculture (Alimba, 2013; Alimba *et al.*, 2006). The safe disposal and effective management of these generated wastes pose significant global challenges. Government initiatives in many countries promote recycling of consumer products to retrieve potentially toxic substances that might otherwise end up in landfills, such as heavy metals found in batteries and electronic devices (Orloff and Falk, 2003; Owonikoko and Alimba, 2024a,b). Nevertheless, antiquated, poor, and often illegal hazardous waste disposal practices continue to have detrimental effects on local communities, particularly in developing countries (Fazzo *et al.*, 2017; Owonikoko and Alimba, 2024a,b). Historical incidents such as mercury poisoning in Minamata Bay, Japan, and the Love Canal incident in the United States serve as stark reminders of the consequences of negligent hazardous waste management (Polprasert, 1996).

The health implications of human exposure to hazardous waste in the environment have garnered increasing attention in recent years from both governmental bodies and the general public. As nations advance industrially and economically, public health organizations pay more attention to environmental health issues pertaining to hazardous wastes (Orloff and Falk, 2003; Alimba, 2013). In Africa, hazardous waste is recognized as one of the top three environmental risk factors (Prüss-Üstinand and Corvalán, 2006), contributing to a significant portion (one-third) of the region's disease burden, with hazardous waste management representing a key concern (Fazzo *et al.*, 2017). Hazardous waste can be characterized based on its toxicity, which may be chronic or acute. The presence of harmful substances, particularly heavy metals such as arsenic (As), mercury (Hg), nickel (Ni), cadmium (Cd), lead (Pb), copper (Cu), and chromium (Cr), in concentrations that exceed regulatory limits, defines the hazardous nature of these wastes (Rosenfeld and Feng, 2011; Alimba *et al.*, 2016a). These metals are recognized for their high potential to induce environmental pollution and biological toxicity (Wu *et al.*, 2024) and have thus been listed among the top 20 most dangerous substances by the U.S. EPA and the Agency for Toxic Substances and Disease Registry (ATSDR) (ATSDR, 2007; Rai *et al.*, 2019). Heavy metals are the primary toxic constituent of industrial hazardous waste, and their contamination of agricultural soils is one of the most pressing environmental issues today.

The safety of food crops grown for human consumption within the vicinity of hazardous waste sites has been a source of public health concern due to metal contamination (Udofia *et al.*, 2020). Metals absorbed by crops can trigger metabolic disorders in consumers, particularly those who consume crops grown in contaminated soils. A notable example is the case of Zamfara State, Nigeria, where lead contamination in the soil surrounding a gold ore mining site exceeded 100,000 mg/kg, prompting the declaration of an unprecedented public health emergency (Moszynski, 2010). As an aftermath of this pollution, lead poisoning took the lives of over 400 people, many of whom were children (the highest metal poisoning mortality rate in Nigerian history) (Galadima and Garba, 2012). According to Tirima *et al.* (2018), the primary exposure route of this incidence was associated with the consumption of Pb-contaminated food crops harvested from the Pb-contaminated soils. Once heavy metals are introduced into the soil, they persist for long periods due to their resistance to biological or chemical degradation (Kasam *et al.*, 2018). Heavy metals, as chemical xenobiotics, have the potential to bioaccumulate in plants and bio-magnify within the food chain, ultimately reaching humans and other larger mammals, where they may cause morbidity and mortality (Alimba *et al.*, 2021; Chukwuka *et al.*, 2023).

Long-term exposure to heavy metals through the consumption of contaminated food crops has been linked to damage to vital organs such as the liver and kidneys, as well as the nervous system (Chukwuka *et al.*, 2023; Wu *et al.*, 2024). In fact, some of the aforementioned heavy metals lack any known biological function and are categorized as "known" or "probable" human carcinogens by the U.S. Environmental Protection Agency (EPA) and the International Agency for Research on Cancer (IARC) (Tchounwou *et al.*, 2012; Chukwuka *et al.*, 2023). This chapter explores the health risks associated with the consumption of food crops contaminated with heavy metals originating from hazardous waste. It discusses the classification of hazardous waste and the common disposal practices that contribute to heavy metal pollution in soils. Additionally, this chapter provides insights into the bioavailability of these chemical xenobiotics in soil, their translocation and bioaccumulation in edible plants, and the mechanisms through which these contaminants induce health hazards in consumers.

2. Classification and Disposal Methods of Hazardous Wastes

Hazardous waste encompasses byproducts from industrial, residential, commercial, and healthcare activities that pose significant risks to both human health and the environment. The U.S. Environmental Protection Agency (EPA) has established hazardous waste listings to specify waste types that are deemed sufficiently dangerous to warrant regulatory oversight. A waste product can be designated as hazardous and included in the list if it is assessed to present a substantial risk (Shukla *et al.*, 2020). This listing system assists industrial waste handlers in identifying hazardous waste (U.S. EPA, 2005) and is essential in determining the optimal management, storage, treatment, and disposal methods to mitigate adverse impacts. Identification of the waste's source is often solely sufficient to establish whether it is listed. The EPA uses four main criteria to classify waste as hazardous (U.S. EPA, 2005):

- **Potential threat without regulation:** Wastes containing harmful chemicals and other factors that suggest they could pose threat to the environment and human health without specific regulation. Such wastes are classified as toxic listed wastes.
- **Inherent hazard despite management:** Wastes containing substances that, even with appropriate management, pose inherent risks to health and the environment. Such wastes are categorized as acutely hazardous wastes.
- **Exhibiting hazardous characteristics:** Wastes that demonstrate one or more of the four hazardous characteristics—ignitability, corrosivity, reactivity, and/or toxicity.
- **Meeting legislative definitions:** When the EPA reasonably determines that a waste meets the broader statutory definition of hazardous waste developed by Congress.

2.1. *Listing categories of hazardous wastes*

On the basis of one or more of the aforementioned listing criteria, over 450 waste types have been formally recognized as hazardous by the EPA and are further classified into specific categories: F-list, K-list, P-list, and U-list (U.S. EPA, 2005; Kanagamani *et al.*, 2021). This classification is based on the characteristics, sources, industrial processes, or generation rate of the hazardous waste (Rosli *et al.*, 2024).

2.1.1. *F-List*

The F-list comprises hazardous wastes from non-specific or generalized sources, reflecting the diverse industrial processes that generate solid waste across sectors. Wastes in this category are assigned EPA waste identification codes beginning with "F" (e.g., F001 and F002) (Shukla *et al.*, 2020). These F-listed wastes are classified according to the type of manufacturing or industrial operation involved, and they include spent halogenated solvents such as carbon tetrachloride and trichloroethylene used in degreasing, metal treatment baths, wastewater treatment sludge from electroplating processes, and dioxin-containing chemicals or precursors (Kanagamani *et al.*, 2021).

2.1.2. *K-List*

Hazardous wastes produced as a result of specific industrial and manufacturing processes are included in the K-list and are therefore regarded as source-specific wastes (U.S. EPA, 2005). K-listed wastes contain chemicals that have been utilized for their intended purpose, and according to the U.S. EPA, they must be generated from processes in the following industrial sectors: wood preservation, organic chemicals manufacturing, pesticide manufacturing, petroleum refining, veterinary pharmaceuticals manufacturing, inorganic pigment manufacturing, inorganic chemicals manufacturing, explosives manufacturing, iron and steel production, primary aluminum production, secondary lead processing, ink formulation, and coking (coal processing to produce coke) (Kanagamani *et al.*, 2021).

2.1.3. *P and U lists*

The P and U lists cover hazardous wastes consisting of unused chemicals in either pure form (100% active ingredient) or commercial-grade form (nearly pure with minor impurities) (Shukla *et al.*, 2020; Kanagamani *et al.*, 2021). Unlike the F and K lists, which catalog wastes from manufacturing processes, the P and U lists are applicable to unused discarded chemicals, whether due to accidental spillage, expiration, or unsuitability for their intended purpose (U.S. EPA, 2005). This includes pesticides and pharmaceuticals.

These listings by the U.S. EPA serve to guide the regulation and disposal practices necessary to protect human health and environmental integrity. A section of the EPA's statutory definition of hazardous waste under the Resource Conservation and Recovery Act (RCRA) states that "it is any substance that may pose substantial threat or potential hazard to human health or the environment when improperly treated, stored, transported, disposed of, or otherwise managed". This suggests that the management of hazardous waste is critical for the bioavailability of its toxic constituents, such as heavy metals, within environmental matrices. As disposal represents the final and most critical phase of any hazardous waste management plan, it becomes imperative to examine common disposal practices that may contribute to metal pollution of the environment. The choice of a disposal method is influenced by factors such as capital investment, operational and maintenance expenses, and technological requirements. Nonetheless, a range of disposal options may be employed within a country to address hazardous waste management needs (Table 1) (Visvanathan, 1996; Shukla *et al.*, 2020).

Table 1. Hazardous waste disposal methods, primary sources, and regulatory frameworks in selected countries.

Country	Primary source(s) of hazardous waste	Disposal method(s) employed	Key regulation(s)/policy frameworks governing hazardous waste management
Brazil	Industrial operations, mining, agriculture	Landfills, incineration, composting	Law No. 12.305 on the National Policy on Solid Waste Management, emphasizing waste reduction, recycling, and safe disposal protocols.
Colombia	Power plants, chemical industry	Landfills	Decree 4741 of 2005; Basel Convention for transboundary movement control and environmentally sound management of hazardous wastes.
Costa Rica	Hospital wastes, chemical industry	Open landfills	Law 8839 of 2012, addressing comprehensive waste management but with ongoing challenges in safe hazardous waste disposal.

(*Continued*)

Table 1. (*Continued*)

Country	Primary source(s) of hazardous waste	Disposal method(s) employed	Key regulation(s)/policy frameworks governing hazardous waste management
Egypt	Agriculture, chemical industry	Landfills	Waste Management Law No. 202 of 2020, providing guidelines for hazardous waste disposal and environmental protection efforts.
Honduras	Hospital wastes, mining, agriculture	Landfills	Agreement No. 1567–2010; Basel Convention, focusing on safe disposal and management of hazardous waste in accordance with international standards.
India	Mining, metallurgy, power plants, chemical industry	Landfills, incineration, deep well disposal	Hazardous Waste Management Rules, establishing stringent controls for waste generation, treatment, and disposal, including tracking and reporting.
*Nigeria	Petrochemical industry, mining, healthcare activities, electronic wastes	Open landfills, incineration	Harmful Waste (Special Criminal Provision, etc.) Act of 1988, criminalizing improper hazardous waste management and focusing on health and environmental safety.
Poland	Mining, electric power plants, metallurgy, agriculture, chemical, and food industries	Landfills, incineration	Waste Act of 2012, aligning with the EU Waste Framework Directive, emphasizing safe disposal and reduction of hazardous waste.
Spain	Manufacturing, mining	Incineration, landfills	National Waste Law (10/1998), with a focus on sustainable waste management practices and compliance with EU regulations.
Taiwan	Electronic, chemical, and metallurgical industries	Incineration, immobilization (solidification), landfilling	Waste Disposal Act; Basel Convention, promoting environmentally sound waste management and cross-border waste transport regulations.

Table 1. (*Continued*)

Country	Primary source(s) of hazardous waste	Disposal method(s) employed	Key regulation(s)/policy frameworks governing hazardous waste management
Trinidad and Tobago	Hospital waste, petrochemical industry	Bio-remediation, incineration stabilization in cement, landfilling	Waste Management Rules, 2021; Waste Management (Fees) Regulation, 2021, introducing specific guidelines for waste treatment, fees, and disposal standards.

Notes: N.B. The regulatory frameworks listed for each country are not exhaustive; additional local regulations may also apply.

Sources: Adapted from Orloff and Falk (2003) and *Zangina and Ali (2021).

2.2. *Disposal methods for hazardous wastes*

2.2.1. *Landfills*

Landfilling, one of the simplest and most practiced hazardous waste disposal methods, is widely favored due to its relatively low technological demands. A landfill serves as a disposal facility where hazardous wastes are placed and stored within the soil, a practice utilized across various nations (Visvanathan, 1996). Conventional landfilling of hazardous waste, such as burial in an unlined, excavated pit or surface dumping on unused land (Orloff and Falk, 2003), can lead to harmful concentrations of toxic substances leaching into the soil, thereby resulting in environmental (soil) pollution. The heavy metal content of leachate, which originates from industrially generated hazardous waste, presents a more severe public health concern than that of other constituents, such as biological agents and organic substances (Kanagamani *et al*., 2021). To prevent the release of hazardous constituents such as heavy metals from solid wastes into the soil, secure landfills that employ barrier systems designed to collect any residual hazardous substances in the waste and prevent the percolation of leachate should be used to manage hazardous wastes (Shukla *et al*., 2020; Kanagamani *et al*., 2021). Nevertheless, many hazardous waste/secure landfills are projected to experience leakage due to outdated designs or faults in current construction and maintenance practices (Rosenfeld and Feng, 2011). Furthermore, as illustrated in Table 1, many countries, particularly those in the developing world, continue to rely on open landfill systems for hazardous waste disposal, thereby perpetuating the issue of soil pollution linked to hazardous waste.

2.2.2. *Incineration*

According to the U.S. EPA, incineration involves burning hazardous materials at a temperature high enough to get rid of contaminants (U.S. EPA, 2012). This thermal oxidation process, typically conducted in an "incinerator," converts hazardous wastes into gaseous emissions and incombustible solid residues through the use of pure oxygen or air (Visvanathan, 1996). Incineration is employed for hazardous wastes that cannot be recycled or safely disposed of in a landfill due to their high toxicity levels or potential risk of disease transmission. Consequently, this disposal practice facilitates activities such as volume and weight reduction, detoxification, and energy recovery (waste to energy) (Rosenfeld and Feng, 2011). The gaseous byproducts are released into the atmosphere, while the toxic solid residues may ultimately be disposed of in landfills. Although incineration effectively destroys various harmful chemicals, including solvents, PCBs (polychlorinated biphenyls), and pesticides, it does not eliminate metal components such as lead and chromium present in the solid residues of hazardous wastes (U.S. EPA, 2012). Instead, these heavy metals undergo oxidation and transformation within incinerator residues, remaining at concentrations roughly equivalent to those in the original waste stream (Rosenfeld and Feng, 2011). For instance, Sedman and Esparza (1991), after assessing 20 incinerators across the United States, found that five of these facilities discharged heavy metals at concentrations posing a serious risk. This limitation suggests that widely utilized incineration may not be an ideal disposal method for mitigating metal pollution from hazardous waste in the environment.

2.2.3. *Deep well disposal or underground injection*

Another alternative method for hazardous waste disposal involves injecting liquid industrial waste into deep underground wells. In this approach, liquid waste, whether treated or untreated, is injected thousands of feet below the soil surface through a reinforced well shaft into porous injection zones enclosed by impermeable rock layers (Rosenfeld and Feng, 2011). Ideally, the geologic formation of these subsurface reservoirs should minimize the risk of contaminant migration (Kanagamani *et al.*, 2021). Although deep-well injection is cost-effective and typically requires minimal waste pretreatment, there remains a risk of hazardous waste leakage, which can lead to environmental pollution, particularly of underground water resources (Kanagamani *et al.*, 2021). For instance, in 1979, it was discovered that pulping liquors disposed of by Hammermill Paper Company were leaking from an abandoned injection well on a small island in Lake Erie, Pennsylvania. This leakage contaminated both Lake Erie, a primary drinking water source, and additional underground water supplies (Gordon and Bloom, 1985).

3. Heavy Metal Leaching and Soil Contamination from Hazardous Wastes

Soil is essential for sustaining terrestrial ecosystems and their biodiversity; however, it also serves as a significant sink for heavy metal pollutants (Das *et al.*, 2023). Regulatory bodies, such as the European Environment Agency (EEA), have established permissible limit values for heavy metals in soil to mitigate pollution risks. These limits include Hg (0.20 mg/kg), Cd (0.44 mg/kg), Pb (0.48 mg/kg), Cr (0.20 mg/kg), and As (0.11 mg/kg) (EU, 2002; Baritz *et al.*, 2023). Despite these regulatory measures, anthropogenic activities, particularly the migration of leachates from improperly managed or disposed hazardous waste, often result in heavy metal concentrations exceeding these thresholds. Notably, arsenic has been identified at 781 of the 1300 hazardous waste sites proposed by the US. EPA for inclusion on the National Priority List (ATSDR, 2000; Tchounwou *et al.*, 2012). The potential environmental risk posed by heavy metals in hazardous waste is closely associated with their leaching toxicity (Wu *et al.*, 2024). The concentrations of heavy metals in soils surrounding waste dumps are influenced by several factors, including the composition and types of waste materials, the area's topography, surface runoff dynamics, and the extent of scavenging activities (Järup, 2003).

Specific heavy metals are prevalent in various consumer and industrial products, such as phone batteries, electronic components, plastics (Cd), wood preservatives, paints, dyes, pharmaceuticals, herbicides (As), and electroplated metal scraps, ceramic materials, and spent Ni–Cd batteries (Ni). The indiscriminate disposal of such waste products at dumpsites significantly increases the release of metal-laden leachates, which infiltrate the underlying soil and exacerbate contamination (Sanga *et al.*, 2024). During decomposition or oxidation processes, heavy metals are mobilized from the waste leachates into the surrounding soil (Nyiramigisha *et al.*, 2021). Given the global challenge of metal-induced soil pollution from hazardous waste mismanagement, it is crucial to develop a comprehensive understanding of the pathways that emphasize the complexity of the health risks posed by metal contamination in food crops, thereby informing mitigation strategies and policy development. These pathways encompass the bioavailability of metals in agricultural soils, their transfer from contaminated soils to crops, and subsequent bioaccumulation within plant tissues.

3.1. *Bioavailability of metals in agricultural soils*

The bioavailability of heavy metals in soils plays a crucial role in determining their environmental behavior and uptake by plants (Ali *et al.*, 2019). Different heavy metals exhibit varying degrees of bioavailability in soils, primarily influenced by their speciation—defined as the different ionic or valence states of the same element (e.g., Fe^{2+}, Fe^{3+}, or Cr^{3+}, and Cr^{6+})—as well as the physicochemical properties of the soil, such as pH, redox potential, organic matter content, clay mineral composition, and oxide concentrations (Mirsal, 2008; Kim *et al.*, 2015; Ali *et al.*, 2019). For instance, hexavalent chromium (Cr^{6+}) is highly soluble in its oxidized state, hence, more bioavailable than its trivalent counterpart (Cr^{3+}) (Jaishankar *et al.*, 2014). Similarly, metal speciation also affects mobility, as seen with arsenic, where trivalent arsenic (As^{3+}) is more mobile in the environment than pentavalent arsenic (As^{5+}) (Mirsal, 2008). At high pH levels, heavy metals are typically adsorbed onto hydroxyl surfaces of oxides or clay minerals, forming insoluble hydroxo-metal complexes. Conversely, at low pH, metals are more likely to exist as free ionic species or soluble organometallic complexes, making them more mobile and bioavailable (Olaniran *et al.*, 2013; Kim *et al.*, 2015). This is because, at low pH, heavy metals are predominantly non-specifically adsorbed onto cation exchange sites of clay minerals and oxides through electrostatic interactions. These interactions are associated with the exchangeable metal fraction, which is considered to be more readily available for environmental processes (Kim *et al.*, 2015).

The process of bioavailability begins with the transition of a metal from the solid phase—where it is largely unavailable for biological or environmental processes—to the liquid phase, where it becomes potentially accessible to soil biota. Upon entering the soil, metals can bind to the solid phase via adsorption onto mineral (e.g., clay, oxides, and hydroxides) or organic (e.g., humic substances) matter. These contaminants are subsequently released into the liquid phase in response to changes in the chemical environment of the soil solution, e.g., reduction in pH levels. Once in the liquid phase, contaminants can migrate through transport mechanisms such as diffusion and dispersion, ultimately coming into contact with the membranes of biological organisms like plants (Gianniantonio *et al.*, 2020). Metals with higher bioavailability in soil are more readily accumulated by plants, thereby exhibiting greater bioaccumulation potential. Consequently, the bioavailability of metals in soil can be effectively estimated by evaluating the extent of heavy metal bioaccumulation in plants (Ali *et al.*, 2019).

3.2. *Heavy metal transfer from soil to plants and bioaccumulation in plant tissues*

The soil-to-plant transfer of heavy metals is a critical process in the trophic transfer of these contaminants through food chains (Ali *et al.*, 2019). The transfer factor (TF), also referred to as the translocation factor, serves as an essential index for assessing the potential of heavy metals to mobilize from soil to plant tissues. It is a key parameter for evaluating the uptake of heavy metals by plants (Mirecki *et al.*, 2015; Osae *et al.*, 2023; Sharafi *et al.*, 2024). The TF is calculated as the ratio of the concentration of a specific heavy metal in plant tissue to its total concentration in the soil extract (Mirecki *et al.*, 2015). Higher TF values (≥ 1) indicate that the plant effectively absorbs metals from the soil, making it more suitable for applications such as phytoextraction and phytoremediation. Conversely, lower TF values (<1) suggest a limited uptake of metals by the plant, rendering it safer for human consumption (Mirecki *et al.*, 2015; Sharafi *et al.*, 2024). The total concentration of heavy metals in the soil is used in TF estimations, as the soil's capacity to store heavy metals significantly influences their mobility from the solid phase to the liquid phase (Sharafi *et al.*, 2024), which ultimately affects their bioavailability.

The terms Bioaccumulation Factor (BAF), Metal Transfer Factor (MTF), Accumulation Factor (AF), and Bioconcentration Factor (BCF) are often used interchangeably with Transfer Factor (TF) to describe similar concepts (Rai *et al.*, 2019). These indices collectively quantify the extent to which heavy metals accumulate in biological organisms (plants) relative to their concentrations in the surrounding environmental medium (soil) (Ali *et al.*, 2019). Heavy metals are absorbed by plants from soil pores in their ionic forms, with the specific form varying depending on the metal (Rai *et al.*, 2019). The availability of metals for plant uptake is influenced by the plants' physiological requirements for micronutrients and their capacity to absorb or exclude toxic elements (Smical *et al.*, 2008). Heavy metal transfer from soil to plant systems is complex; however, the roots play the most important role in the processes of uptake and translocation. Bioavailable metals are adsorbed onto the root surface and subsequently transported across the cellular membrane into the root apoplast, where they accumulate. The translocation of metal cations from root cells to the xylem and phloem tissues can occur via two primary pathways: the apoplastic pathway, involving passive diffusion, or the symplastic pathway, which is facilitated by active transport mechanisms mediated by metal ion carriers or complexing agents (Yan *et al.*, 2020; Vanisree *et al.*, 2022). Through these pathways, heavy metals are transported and distributed to various tissues and organs within the plant, including those that constitute its edible parts, such as leaves, fruits, and grains.

In addition to metal speciation and soil physicochemical properties, the availability of metals for uptake is strongly contingent on plant species and their adaptive mechanisms to environmental conditions (Mirecki *et al.*, 2015; Sharafi *et al.*, 2024). For instance, herbs typically absorb lower quantities of metals compared to fast-growing plants, such as lettuce, spinach, carrot, and tobacco (Smical *et al.*, 2008). The distribution of heavy metals within plant tissues also varies; roots and leaves tend to accumulate the highest concentrations, while fruits and seeds exhibit the lowest (Mirecki *et al.*, 2015). This observation aligns with numerous past studies showing that leafy vegetables generally exhibit the highest BCF/TF values, followed by tuberous plants, with the lowest values found in horticultural crops and fruits (García *et al.*, 2009; Melgar *et al.*, 2009; Liu *et al.*, 2012; Chang *et al.*, 2014; Yang *et al.*, 2018). Table 2, which presents the concentrations of specific heavy metals in food crops globally, further underscores the bioaccumulation potential of plants in soils impacted by metal contamination. The bioaccumulation of heavy metals in biota, such as plants, is a critical concern from environmental and public health perspectives, with significant implications for both wildlife and human health (Ali *et al.*, 2019).

Table 2. Heavy metal concentrations (mg/kg) in globally represented food crops, highlighting bioaccumulation trends.

Food crops (cereals, fruits, vegetables)	Country of investigation	Pb	Cd	Ni	Cr	As	Cu	References
Onions, tomato, carrot, maize, cucumber, cabbage	Nigeria	2.62–20.49*	0.18–0.58*	NA	0.75–1.81	NA	8.91–13.13	Obiora *et al.* (2016) and Bawa 2023
Spinach, coriander, onion, tomato	India	1.08–3.1*	0.17*	0.73–3.2	2.9*	NA	0.09–5.65	Chary *et al.* (2008) and Gupta *et al.* (2022)
Wheat, tomato, garlic, radish, carrot	Pakistan	0.91–3.96*	NA	NA	>0.18	NA	NA	Khan *et al.* (2013)
Lettuce	United States of America	NA	NA	NA	NA	27.3	NA	de Oliveira *et al.* (2017)
Chinese cabbage, Brassica spp., maize, radish, cauliflower	China	0.013–0.02	0.12–1.70*	NA	0.08–0.38	NA	0.16–0.85	Khan *et al.* (2008) and Junhe *et al.* (2017)
Potato	Egypt	0.08	0.02	NA	0.83	NA	NA	El-Kady and Abdel-Wahhab (2018)
Rice, olive, guava, banana, papaya	Bangladesh	1.09*	0.08	2.62	2.33*	NA	14.7	Hasan *et al.* (2020)

(*Continued*)

Table 2. *(Continued)*

Food crops (cereals, fruits, vegetables)	Country of investigation	Pb	Cd	Ni	Cr	As	Cu	References
Lettuce, tomato, potato, carrot, beans	Romania	0.08–2.22*	0.04–1.52*	NA	NA	NA	6.0–46.0*	Heghedűş-Mîndru *et al.* (2023)
Apple	Kosovo	1.49–2.17*	0.17–0.21*	15.33–17.2*	8.22–11.6*	0.009	1.05–2.87	Imeri *et al.* (2019)
Apple	Ukraine	1.35–3.89*	0.088–0.28*	NA	0.023–0.25	NA	0.77–1.26	Medvedeva *et al.* (2021)
Tomato, zucchini	Spain	0.22	0.20*	0.52	0.28	BDL	6.58	Antisari *et al.* (2015)
Millet, tomato, watermelon, bell pepper, melon	Uzbekistan	0.06–0.49*	0.031–0.091	0.03–3.85	0.01–2.21	0.57–8.72	2.24–30.64*	Uzakov and Uzakov (2023)
Cocoyam	Tanzania, Uganda, Kenya	2.33*	NA	NA	NA	0.93	5.22	Mongi and Chove (2020)
Amaranth, eggplant	Ghana	21.13*	0.35*	3.56	6.72*	0.64	16.27	Osae *et al.* (2023)
Permissible limit(s)		**0.3**[a]	**0.1–0.2**[a]	**10**[a]**, 67**[b]	**2.3**[a]	**30**[a]	**20**[a]**, 73**[b]	FAO/WHO (2001)[a] and FAO/WHO (2011)[b]

Notes: NA: Not available (according to authors' published reports), BDL: Below Detection Limit.
* represents metal levels in food crops exceeding one or both of the established safety thresholds.

4. Health Risks Associated With the Consumption of Metal-Laden Food Crops

Growing food crops in heavy metal-contaminated environments, such as agricultural soils near hazardous waste sites, has been documented to result in the bioaccumulation of these toxic elements within the human food chain (Rai *et al.*, 2019; Munir *et al.*, 2022; Sanga *et al.*, 2024). The ingestion of such contaminated crops represents a major pathway for metal-induced toxicity in humans and other mammals (Fig. 1). Orisakwe *et al.* (2015) reported that plant-based foods account for approximately 50% of total dietary intake of lead (Pb), cadmium (Cd), and mercury (Hg). Particularly, rice, a staple food in many regions of the globe, has been identified as a significant source of Cd and Pb in human consumers (Ihedioha *et al.*, 2016). For instance, between 1910 and 2007, the consumption of Cd-contaminated rice cultivated in the Jinzu River basin, Toyama, Japan, led to Itai-itai disease, a severe form of cadmium poisoning. This tragic incident resulted in over 100 fatalities and left many individuals with debilitating deformities (Matsunami, 2010). The biomagnification of trace metals within organisms at higher trophic levels can lead to elevated concentrations, posing significant health risks to both

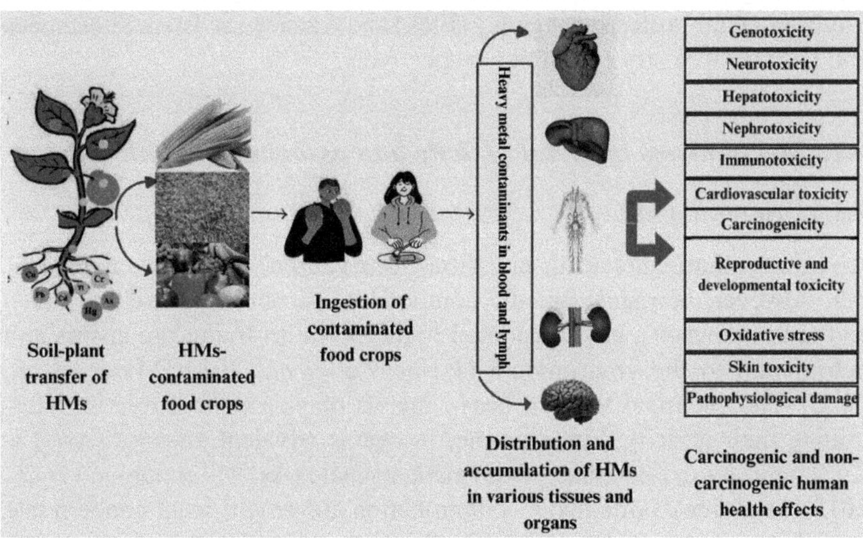

Fig. 1. Schematic illustration of heavy metal (HM) uptake by food crops, oral exposure pathway, and associated toxicological impacts on human health.

these organisms and their human consumers. Consequently, higher trophic-level organisms in the food chain are at greater risk of adverse effects (Ali *et al.*, 2019).

While trace metals such as zinc (Zn), iron (Fe), copper (Cu), manganese (Mn), and nickel (Ni) are essential for the physiological functions of living organisms in trace amounts (Bawa, 2023), elevated and prolonged exposure through contaminated food can induce toxicity and disrupt critical biological processes. For several of these metals, the threshold between essentiality and toxicity is notably narrow (Ali *et al.*, 2019). Conversely, non-essential metals, including lead (Pb), cadmium (Cd), arsenic (As), and chromium (Cr), pose significant health hazards even at minimal concentrations (Alimba *et al.*, 2016b; 2018; Chukwuka *et al.*, 2023). Health hazard indices, including the hazard quotient (HQ), hazard index (HI), and incremental lifetime cancer risk (ILCR), are widely utilized to assess both carcinogenic and non-carcinogenic health risks associated with the dietary consumption of food crops contaminated with heavy metals (Chary *et al.*, 2008; Chang *et al.*, 2014; Obiora *et al.*, 2016; Udofia *et al.*, 2020; Bawa, 2023; Chukwuka *et al.*, 2023). This section provides a comprehensive analysis of the toxicological effects and potential mechanisms of toxicity of selected heavy metals included among the top twenty toxic chemicals and substance priority list by the U.S. Environmental Protection Agency (EPA) and Agency for Toxic Substances and Disease Registry (ATSDR), respectively.

4.1. *Mechanisms of metal toxicity and associated health impacts*

4.1.1. *Arsenic (As)*

Arsenic contaminants occur in various forms across environmental matrices; however, inorganic arsenic compounds, particularly arsenite (As^{3+}) and arsenate (As^{5+}), are considered highly toxic to living organisms and detrimental to the environment (Jaishankar *et al.*, 2014; Munir *et al.*, 2022). The chemical form of heavy metals plays a critical role in determining their toxicity. For example, inorganic trivalent arsenite (As^{3+}) is 2–10 times more toxic than pentavalent arsenate (As^{5+}) (Tchounwou *et al.*, 2012). High levels of arsenic contamination are a significant concern due to its wide range of adverse health effects. A review by Tchounwou *et al.* (2003) established a strong correlation between arsenic exposure and increased risks of carcinogenic and systemic health effects. Arsenic exposure impacts virtually all organ systems, including the cardiovascular, dermatologic, nervous, hepatobiliary, renal, gastrointestinal, and respiratory systems (Fig. 1; Tchounwou *et al.*, 2003). This widespread toxic effect on different organ systems is attributed to As^{3+}'s ability to interact with thiol or sulfhydryl groups on proteins and enzymes, leading to the impairment of enzymatic function, disruption of mitosis, and interference with cellular respiration. On the other hand, As^{5+} interferes with many biochemical pathways by replacing phosphate (Hughes, 2002). These actions classify arsenic as a potent protoplasmic poison (Jaishankar *et al.*, 2014; Munir *et al.*, 2022).

In humans and some other life forms such as bacteria and fungi, inorganic arsenic compounds, such as arsenic trioxide, are biotransformed through methylation into two primary metabolites: monomethylarsonic acid (MMA) and dimethylarsinic acid (DMA). These metabolites are excreted via urine and serve as reliable biomarkers of chronic arsenic exposure (Hughes, 2002; Tchounwou *et al.*, 2012; Jaishankar *et al.*, 2014). While earlier studies suggested that arsenic methylation functions as a detoxification mechanism, more recent research has demonstrated that certain methylated metabolites, such as trivalent monomethylarsonic acid (MMA^{3+}), exhibit heightened toxicity compared to arsenite (As^{3+}) itself (Tchounwou *et al.*, 2003). Singh *et al.* (2007) reported that MMA^{3+}, an intermediate product of arsenic methylation, is not excreted via urine but remains within cells, where it has the potential to induce carcinogenesis. Compared to inorganic arsenic, enzymatically formed arsenic metabolites within biological systems can cause significant side effects (Munir *et al.*, 2022), further emphasizing their toxicological importance.

The genotoxicity of arsenic results in various DNA alterations, including chromosomal abnormalities, mutations, micronuclei formation, deletions, and sister chromatid exchanges (Tchounwou *et al.*, 2012; Munir *et al.*, 2022; Mitra *et al.*, 2022). Although arsenic is considered a weak mutagen due to its lack of direct interaction with DNA (Mitra *et al.*, 2022), its genotoxic effects are primarily mediated through mechanisms such as oxidative stress induction (Fig. 2) and disruption of DNA repair processes (Pierce *et al.*, 2012). Arsenic poisoning (arsenicosis) also impacts the central nervous system, leading to cognitive impairments and alterations in synaptic transmission and neurotransmitter balance. These effects are facilitated by caspase-induced apoptosis in neural cells via activation of the MAPK signaling pathways (Garza-Lombó *et al.*, 2019). Additionally, arsenic acts as a carcinogen by inducing DNA methylation and damage

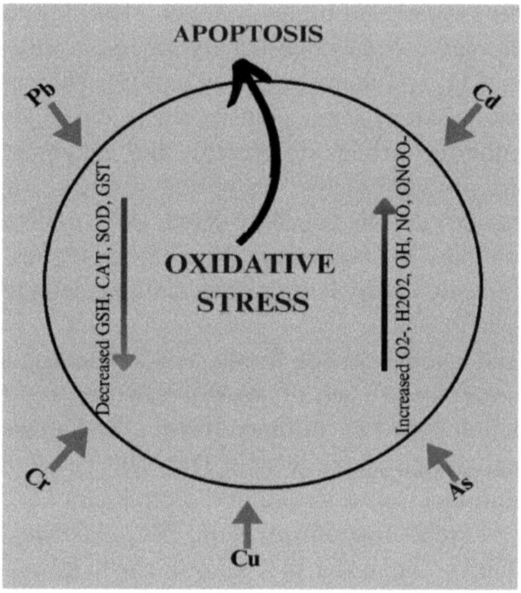

Fig. 2. Oxidative stress and apoptotic cell death characterized by imbalance of antioxidant levels (decrease) and ROS generation (increase) in living cells; a common mechanism of metal-induced toxicity (GSH – glutathione, CAT – catalase, SOD – superoxide dismutase, GST – glutathione S-transferases, O_2 – superoxide, H_2O_2 – hydrogen peroxide, OH^- – hydroxyl, NO – nitric oxide, $ONOO^-$ – peroxynitrite).

(Mitra *et al.*, 2022). It further increases carcinogenic risk by binding to DNA-associated proteins and inhibiting DNA repair processes (Garcia-Esquinas *et al.*, 2013). Beyond its carcinogenic effects, inorganic arsenic has been implicated in reproductive toxicity. In males, it inhibits sperm production in the testes and disrupts testosterone and gonadotropin levels (Kim and Kim, 2015). In females, arsenic interferes with endometrial angiogenesis, a process critical for embryo development during pregnancy (Milton *et al.*, 2017).

4.1.2. *Lead* (*Pb*)

Lead is one of the most systemic toxicants to the human body, affecting multiple organ systems (Fig. 1) (ATSDR, 1999). The majority of absorbed lead biomagnifies in the kidneys, followed by the liver and other soft tissues, such as the heart and brain (Flora *et al.*, 2006). The kidney is particularly vulnerable, with Pb exposure potentially resulting in acute or chronic nephropathy. These conditions are characterized by proximal tubular dysfunction, hyperplasia, interstitial fibrosis, tubular atrophy, renal failure, and glomerulonephritis (Mitra *et al.*, 2022). The nervous system is the most sensitive target of lead poisoning. Early neurological symptoms include headache, reduced attention span, irritability, memory loss, and cognitive dullness (ATSDR, 1999). Lead also exhibits carcinogenic properties, disrupting DNA repair mechanisms, cellular tumor-regulating genes, and chromosomal structure through the release of reactive oxygen species (ROS). It further interferes with transcription by displacing zinc in certain regulatory proteins (Mitra *et al.*, 2022).

Chronic lead exposure increases oxidative stress in hepatocytes, leading to liver damage. This hepatotoxicity manifests as glycogen depletion and cellular infiltration, which can progress to chronic cirrhosis (Hegazy and Fouad, 2014). Additionally, lead exposure can impair immune responses (cell-mediated and humoral) by altering T-cells' function, thereby increasing susceptibility to autoimmunity, hypersensitivity, and cancer (Mishra, 2009). Cardiovascular disorders (CVDs), including arteriosclerosis, hypertension, thrombosis, and atherosclerosis, are also associated with chronic lead exposure. These effects are mediated through mechanisms such as increased ROS production (Fig. 2), reduced nitric oxide (NO) bioavailability, elevated vasoconstrictor prostaglandins, and dysregulation of the renin-angiotensin system (Mitra *et al.*, 2022).

Lead exerts its toxic effects in living cells through two primary mechanisms: ionic mimicry and oxidative stress (Jaishankar *et al.*, 2014; Munir *et al.*, 2022). Exposure to elevated levels of lead increases ROS production while depleting antioxidant defenses, such as GSH, CAT, and SOD (Fig. 2), which are crucial for detoxifying reactive intermediates and repairing cellular damage. Excessive ROS accumulation disrupts the structural integrity of cells, damaging nucleic acids, proteins, lipids, and membranes, ultimately leading to oxidative stress (Flora *et al.*, 2008; Munir *et al.*, 2022). The ionic mechanism of lead toxicity arises from the ability of Pb^{2+} to mimic and displace essential cations, such as Ca^{2+}, Mg^{2+}, Fe^{2+}, and Na^{+}. This displacement disrupts key cellular metabolic processes, causing significant alterations in cell adhesion, intracellular and intercellular signaling, protein folding and maturation, ionic transport, apoptosis, enzyme regulation, and neurotransmitter release. Additionally, Pb^{2+} interference affects the expression of calcium-dependent proteins such as calmodulin and protein kinase C, which are critical for neural excitation and memory storage (Flora *et al.*, 2008; Tchounwou *et al.*, 2012; Jaishankar *et al.*, 2014).

4.1.3. *Cadmium (Cd)*

According to the ATSDR ranking, cadmium (Cd) is classified as the seventh most toxic heavy metal. This toxicant affects a range of organ systems (Fig. 1), with the liver and kidneys being particularly sensitive targets (Munir *et al.*, 2022). Chronic exposure to cadmium is associated with renal tubular acidosis, renal failure, and hypercalciuria (Friberg *et al.*, 2019), often accompanied by pronounced nephrotoxic symptoms, such as glucosuria, Fanconi-like syndrome, phosphaturia, and aminoaciduria (Mitra *et al.*, 2022). Cadmium readily accumulates in the liver, inducing oxidative stress and subsequent hepatocellular damage, which contributes to various hepatic dysfunctions (Mitra *et al.*, 2022). The circulatory system also serves as a significant route for cadmium distribution, with blood vessels being primary targets of its toxicity (Tchounwou *et al.*, 2012). Even at low concentrations, cadmium poses substantial cardiovascular risks, including hypertension (Tellez-Plaza *et al.*, 2008), carotid atherosclerosis (Messner *et al.*, 2009), peripheral arterial disease (Navas-Acien *et al.*, 2004), myocardial infarction (Everett and Frithsen, 2008), and heart failure (Peters *et al.*, 2010). Furthermore, studies by Ma *et al.* (2021) and Pouillot *et al.* (2022) implicated cadmium toxicity in osteoporosis and bone fractures, demonstrating its ability to upregulate dynamin-related protein.

Cadmium poisoning has been associated with neurodegenerative disorders, including amyotrophic lateral sclerosis, multiple sclerosis, Parkinson's disease, Alzheimer's disease, and Huntington's disease. These conditions are often accompanied by memory loss and behavioral changes (Rahimzadeh *et al.*, 2017; Mitra *et al.*, 2022). Prolonged exposure to elevated cadmium levels is also linked to pulmonary dysfunction (Das *et al.*, 2023). The International Agency for Research on Cancer (IARC) has classified cadmium as a human carcinogen, majorly based on consistent evidence linking occupational cadmium exposure to an increased risk of lung cancer. This classification is further supported by animal studies, particularly in rodents, which identify the pulmonary system as a primary target of cadmium toxicity (IARC, 1993). These findings establish the lungs as a definitive site of cadmium-induced human carcinogenesis.

While the precise mechanisms underlying cadmium toxicity remain ambiguous, its roles in organ toxicosis, carcinogenesis, and apoptosis are largely attributed to the enhanced generation of ROS (Fig. 2; Tchounwou *et al.*, 2012; Rahimzadeh *et al.*, 2017). Additional pathological mechanisms include epigenetic alterations affecting DNA expression, inhibition of heme synthesis, depletion of reduced glutathione, and mitochondrial dysfunction, particularly impairments in cellular respiration and oxidative phosphorylation (Bernhoft, 2013; Rahimzadeh *et al.*, 2017). Cadmium also disrupts cellular homeostasis by competitively interfering with the physiological functions of zinc (Zn) in metallothionein. This interference inhibits the cysteine-rich protein's ability to act as an intracellular free radical scavenger, thereby exacerbating oxidative stress (Jaishankar *et al.*, 2014).

4.1.4. *Chromium (Cr)*

The health hazards associated with chromium exposure are heavily influenced by its oxidation state, ranging from the relatively low toxicity of the elemental form (Cr^0) to the high toxicity of the hexavalent form (Cr^{6+}) (Tchounwou *et al.*, 2012). Among the various oxidation states of chromium present in the environment, trivalent chromium (Cr^{3+}) and hexavalent chromium (Cr^{6+}) are the most commonly encountered due to their ionic stability (Jaishankar *et al.*, 2014). However, given that Cr^{3+} is not readily absorbed through any exposure route or membrane, the toxicity of chromium is primarily attributed to its hexavalent form, Cr^{6+}. Chronic exposure to Cr^{6+} has been recognized as highly toxic to humans, with elevated levels causing damage to the kidneys, liver, immune system, blood, and gastrointestinal tract (Fig. 1; Munir *et al.*, 2022). Cr^{6+}-induced hepatotoxicity has been associated with histopathological alterations, including hepatocyte steatosis, parenchymatous degeneration, and necrosis. These toxic effects are primarily mediated by increased ROS production (Fig. 2), lipid peroxidation, suppression of DNA, RNA, and protein synthesis, DNA damage, decreased antioxidant enzyme activity, mitochondrial dysfunction, and apoptosis (Hasanein and Emamjomeh, 2019).

High doses of Cr^{6+} have been shown to compromise immune function, reducing the phagocytic activity of alveolar macrophages and weakening humoral immune responses in Wistar rats (Glaser *et al.*, 1985). Similarly, the development of allergic contact dermatitis was attributed to the immunosuppressive effects of Cr^{6+}in high concentrations (Ahrensboll-Friis *et al.*, 2022). Furthermore, Cr^{6+} has been implicated in inducing structural changes to genetic material, including DNA-protein crosslinks, inter-strand crosslinks, and nucleotide strand breakage in both living organisms and cultured cells (Fang *et al.*, 2014). Cr^{6+}, a potent oxidizing and highly permeant ion, can be reduced by biological reductants, such as glutathione (GSH), ascorbate, thiols, and glutathione reductase. The interaction between Cr^{6+} and these reductants generates reactive intermediates, including superoxide ions, hydrogen peroxide, hydroxyl radicals, and transient chromium species such as Cr^{5+} and Cr^{4+}, along with thiyl radicals (RS). These reactive intermediates can induce oxidative stress within cells (Fig. 2), leading to damage of critical biomolecules, including DNA, proteins, and membrane lipids (Tchounwou *et al.*, 2012; Jaishankar *et al.*, 2014).

4.1.5. *Mercury* (*Hg*)

Mercury is ranked by the ATSDR as the most toxic heavy metal in the environment. It is unique among heavy metals because it occurs in three distinct forms in nature: elemental, inorganic, and organic, each with its own degree of toxicity (Tchounwou *et al.*, 2012). The nervous system is particularly sensitive to all forms of mercury, and elevated exposure to any of them can impair brain function. Symptoms include shyness, tremors, amnesia, irritability, and altered vision or hearing (Jaishankar *et al.*, 2014). Mercury also exerts significant nephrotoxic effects, inducing acute tubular necrosis during acute exposure and causing chronic damage to the epithelium and necrosis in the pars recta of the proximal tubule. These nephrotoxic effects are associated with symptoms, such as acute dyspnea, abdominal pain, tubular failure, and elevated urinary excretion of albumin and retinol-binding protein (Lentini *et al.*, 2017).

The peroxidative activity of mercury generates reactive oxygen species (ROS), contributing to carcinogenesis by damaging biomolecules, such as DNA and proteins. Mercury-induced ROS production also facilitates protumorigenic signaling, promoting the growth of cancerous cells (Reczek and Chandel, 2017; Mitra *et al.*, 2022). Mercury poisoning is a leading cause of dermatological issues. It has been associated with acrodynia, also known as pink disease, which is characterized by a pinkish discoloration of the skin and heightened sensitivity of the palms and soles (Horowitz *et al.*, 2002). Additionally, mercury exposure can result in acute contact dermatitis symptomatized by moderate swelling, scaling, vesiculation, and irritation (Boyd *et al.*, 2000).

Methylmercury (MeHg) is a potent neurotoxic compound known to cause microtubule destruction, mitochondrial damage, lipid peroxidation, and the accumulation of neurotoxic molecules, such as serotonin, aspartate, and glutamate (Patrick, 2002). Once inside the cell, both MeHg and Hg^{2+} induce oxidative stress (Fig. 2) by forming covalent bonds with sulfhydryl and selenohydryl groups in proteins. This interaction disrupts the structure and function of key enzymes involved in cellular repair processes and anti-oxidant defense mechanisms, ultimately damaging cells (Tchounwou *et al.*, 2012). Mercury further interferes with transcription and translation processes, leading to the loss of ribosomes, degradation of the endoplasmic reticulum, and suppression of natural killer cell activity (Jaishankar *et al.*, 2014). The toxic effects of mercury are largely attributed to its chemical reactivity and biological characteristics, with oxidative stress playing a central role in its mechanism of toxicity (Valko *et al.*, 2005).

5. Conclusion

The safety of food crops is a critical global priority for protecting human health, yet they are increasingly threatened by anthropogenic sources of heavy metals, such as hazardous waste. The generation of hazardous waste is a worldwide issue, with the volume of waste produced closely linked to the level of industrialization. The safe disposal and effective management of these wastes present significant challenges globally due to their hazardous composition, particularly the presence of toxic heavy metals. Heavy metals leach into soils, leading to pollution and creating a reservoir of bioavailable contaminants for plant uptake. Plant roots absorb both essential and non-essential heavy metals in solution, which subsequently accumulate in edible plant parts, such as fruits, leaves, roots, and seeds. The ingestion of such contaminated crops by tertiary consumers, including humans, facilitates biomagnification and increases the risk of both carcinogenic and non-carcinogenic health effects. Heavy metals induce biological toxicity through multifaceted mechanisms that affect multiple organs and systems, including the renal, hepatic, cardiovascular, nervous, circulatory, immune, reproductive, and respiratory systems. This chapter enhances understanding of the pathways and impacts of metal-induced toxicity on human and wildlife consumers of contaminated food crops, thereby providing critical insights for implementing and enforcing effective regulations and mitigation strategies. Such efforts are essential for reducing hazardous waste-related metal pollution in agricultural soils and ensuring the safety and sustainability of food production systems.

Bibliography

Agency for Toxic Substances and Disease Registry (ATSDR). (1999). Toxicological Profile for Lead. Public Health Service. U.S. Department of Health and Human Services, Atlanta.

Agency for Toxic Substances and Disease Registry (ATSDR). (2000). Toxicological Profile for Arsenic TP-92/09. Center for Disease Control, Atlanta.

Agency for Toxic Substances and Disease Registry (ATSDR). (2007). Toxicological Profile for Barium. U.S. Department of Health and Human Services, Public Health Service, Atlanta.

Ahrensboll-Friis, U., Simonsen, A.B., Dahlin, J., Isaksson, M., Zachariae, C., and Johansen, J.D. (2022). Allergic contact dermatitis from dyes used in the temple of spectacles. *Contact Dermatitis*, 86, 25–28.

Ali, H., Khan, E., and Ilahi, I. (2019). Environmental chemistry and ecotoxicology of hazardous heavy metals: Environmental persistence, toxicity, and bioaccumulation. *Journal of Chemistry*, Article ID: 6730305. https://doi.org/10.1155/2019/6730305.

Alimba, C.G. (2013). *DNA and Systemic Damage Induced by Landfill Leachates and Health Impacts of Human Exposure to Landfills in Lagos and Ibadan, Nigeria*. Ph.D thesis, Department of Zoology, University of Ibadan, Nigeria, p. 258.

Alimba, C.G. (2024). Alterations in viscera histoarchitecture and organosomatic index as biomarkers of toxicity induced by Aba-Eku and Olusosun solid waste landfill leachates in *Rattus norvegicus*. *Environmental Analysis Health and Toxicology*, 39(2), Article ID: e2024022, 17 pages. https://doi.org/10.5620/eaht.2024022.

Alimba, C.G. and Bakare, A.A. (2006). Municipal landfill leachate induced chromosome aberration in rat bone marrow cells. *African Journal of Biotechnology*, 5, 2053–2057.

Alimba, C.G., Adewumi, O.O., Binuyo, O.M., and Odeigah, P.G.C. (2021). Wild black rats (*Rattus rattus* Linnaeus, 1758) as zoomonitors of genotoxicity and systemic toxicity induced by hazardous emissions from Abule Egba unsanitary landfill, Lagos, Nigeria. *Environmental Science and Pollution Research*, 28, 10603–10621. https://doi.org/10.1007/s11356-020-11325-7.

Alimba, C.G., Aladeyelu, A.M., Nwabisi, I.A., and Bakare, A.A. (2018). Micronucleus cytome assay in the differential assessment of cytotoxicity and genotoxicity of cadmium and lead in *Amietophrynus regularis*. *EXCLI*, 17, 89–101.

Alimba, C.G., Gandhi, D., Sivanesan, S., Bhanarkar, M.D., Naoghare, P.K., Bakare, A.A., and Krishnamurthi, K. (2016a). Chemical characterization of simulated landfill soil leachates from Nigeria and India and their cytotoxicity and DNA damage inductions on three human cell lines. *Chemosphere*, 164, 469–479. https://doi.org/10.1016/j.chemosphere.2016.08.093.

Alimba, C.G., Dhillon, V., Bakare, A.A., and Fenech, M. (2016b). Genotoxicity and cytotoxicity of chromium, copper, manganese, and lead, and their mixture in WIL2-NS human B lymphoblastoid cells is enhanced by folate depletion. *Mutation Research*, 798–799, 35–47.

Antisari, L.V., Orsini, F., Marchetti, L., Vianello, G., and Gianquinto, G. (2015). Heavy metal accumulation in vegetables grown in urban gardens. *Agronomy for Sustainable Development*, 35, 1139–1147. https://doi.org/10.1007/s13593-015-0308-z.

Baritz, R., Amelung, W., *et al.* (2023). Soil monitoring in Europe — indicators and thresholds for soil health assessments. EEA Reports. Publications Office of the European Union, p. 8.

Basel Convention. (1998). Total amount of hazardous wastes and other wastes generated in 1998 (as reported by parties). http://www.basel.int//pub/TABLE1.pdf. Accessed 20 October 2024.

Bawa, U. (2023). Heavy metals concentration in food crops irrigated with pesticides and their associated human health risks in Paki, Kaduna State, Nigeria. *Cogent Food & Agriculture*, 9(1). https://doi.org/10.1080/23311932.2023.2191889.

Bernhoft, R.A. (2013). Cadmium toxicity and treatment. *The Scientific World Journal*, 394652. https://doi.org/10.1155/2013/394652.

Boyd, A.S., Seger, D., Vannucci, S., Langley, M., Abraham, J.L., and King, L.E. (2000). Mercury exposure and cutaneous disease. *Journal of American Academy of Dermatology*, 43, 81–90. https://doi.org/10.1067/mjd.2000.106360.

Chang, C.Y., Yu, H.Y., Chen, J.J., Li, F.B., and Zhang, H.H. (2014). Accumulation of heavy metals in leaf vegetables from agricultural soils and associated potential health risks in the Pearl River Delta, South China. *Environmental Monitoring and Assessment*, 186, 1547–1560.

Chary, N.S., Kamala, C.T., and Raj, D.S.S. (2008). Assessing risk of heavy metals from consuming food grown on sewage irrigated soils and food chain transfer. *Ecotoxicology and Environmental Safety*, 69, 513–524.

Chukwuka, K.S., Adesida, S.O., and Alimba, C.G. (2023). Carcinogenic and non-carcinogenic risk assessment of consuming metal-laden wild mushrooms in Nigeria: Analyses from field based and systematic review studies. *Environmental Analysis, Health and Toxicology*, 38(2), e2023013. https://doi.org/10.5620/eaht.2023013.

Das, S., Sultana, K.W., Ndhlala, A.R., Mondal, M., and Chandra, I. (2023). Heavy metal pollution in the environment and its impact on health: Exploring

green technology for remediation. *Environmental Health Insights*, 17, 11786302231201259. https://doi.org/10.1177/11786302231201259.
de Oliveira, L.M., Suchismita, D., Gress, J., Rathinasabapathi, B., and Chen, Y. (2017). Arsenic uptake by lettuce from As-contaminated soil remediated with *Pteris vittata* and organic amendment. *Chemosphere*, 176, 249–254.
Duan, H., Huang, Q., Wang, Q., Zhou, B., and Li, J. (2008). Hazardous waste generation and management in China: A review. *Journal of Hazardous Materials*, 158, 221–227. https://doi.org/10.1016/j.jhazmat.2008.01.106.
El-Kady, A.A. and Abdel-Wahhab, M.A. (2018). Occurrence of trace metals in foodstuffs and their health impact. *Trends in Food Science and Technology*, 75, 36–45.
European Union (EU). (2002). Heavy metals in wastes. European Commission on Environment. European Commission DG ENV. E3.
Everett, C.J. and Frithsen, I.L. (2008). Association of urinary cadmium and myocardial infarction. *Environmental Research*, 106, 284–286. https://doi.org/10.1016/j.envres.2007.10.009.
Fang, Z., Zhao, M., Zhen, H., Chen, L., Shi, P., Huang, Z., and Shi, X. (2014). Genotoxicity of tri- and hexavalent chromium compounds in vivo and their modes of action on DNA damage in vitro. *PLoS ONE*, 9(8), e103194. https://doi.org/10.1371/journal.pone.0103194.
FAO/WHO. (2001). Food additives and contaminants. Joint Codex Alimentarius Commission, FAO/WHO Food Standards Programme; ALINORM 01/12A.18.
FAO/WHO. (2011). Joint FAO/WHO food standards programme codex committee on contaminants in foods fifth session. The Hague, The Netherlands, March 2011.
Fazzo, L., Minichilli, F., Santoro, M., Ceccarini, A., Della Seta, M., Bianchi, F., Comba, P., and Martuzzi, M. (2017). Hazardous waste and health impact: A systematic review of the scientific literature. *Environmental Health*, 16, 107. https://doi.org/10.1186/s12940-017-0311-8.
Flora, S.J.S., Flora, G.J.S., and Saxena, G. (2006). Environmental occurrence, health effects and management of lead poisoning. In: Cascas, S.B. and Sordo, J. (Eds.), *Lead: Chemistry, Analytical Aspects, Environmental Impacts and Health Effects*. Elsevier Publication, Netherlands, pp. 158–228.
Flora, S.J.S., Mittal, M., and Mehta, A. (2008). Heavy metal induced oxidative stress & its possible reversal by chelation therapy. *Indian Journal of Medical Research*, 128, 501.
Friberg, L., Kjellström, T., Elinder, C.G., and Nordberg, G.F. (2019). Cadmium and health: A toxicological and epidemiological appraisal. https://doi.org/10.1201/9780429260599.
Galadima, A. and Garba, Z.N. (2012). Heavy metals pollution in Nigeria: Causes and consequences. *Elixir Pollution*, 45, 7917–7922.

Garcia, M.A., Alonso, J., and Melgar, M.J. (2009). Lead in edible mushrooms: Levels and bioaccumulation factors. *Journal of Hazardous Material*, 167, 777–783.

Garcia-Esquinas, E., Pollán, M., Umans, J.G., Francesconi, K.A., Goessler, W., Guallar, E., Howard, B.V., Yeh, J., Best, L., and Navas-Acien, A. (2013). Arsenic exposure and cancer mortality in a US-based prospective cohort: The Strong Heart Study. ISEE Conference Abstract, 1, 3037.

Garza-Lombó, C., Pappa, A., Panayiotidis, M.I., Gonsebatt, M.E., and Franco, R. (2019). Arsenic-induced neurotoxicity: A mechanistic appraisal. *Journal of Biological Inorganic Chemistry*, 24, 1305–1316. https://doi.org/10.1007/s00775-019-01740-8.

Gianniantonio, P., Francesca, P., and Irene, R. (2020). Bioavailability and bioaccessibility in soil: A short review and a case study. *AIMS Environmental Science*, 7, 208–225. doi:10.3934/environsci.2020013.

Glaser, U., Hochrainer, D., Kloppel, H., and Kuhnen, H. (1985). Low level chromium (VI) inhalation effects on alveolar macrophages and immune functions in Wistar rats. *Archives of Toxicology*, 57, 250–256. https://doi.org/10.1007/BF00324787.

Gordon, W. and Bloom, J. (1985). Limits to underground injection as a hazardous waste disposal method. *Hazardous Waste & Hazardous Materials*, 2, 411–451.

Gupta, N., Yadav, K., Kumar, V., Prasad, S., Cabral, P.M., Byong-Hun, J., Kumar, S., Abdellattif, M., and Alsukaibi, A. (2022). Investigation of heavy metal accumulation in vegetables and health risk to humans from their consumption. *Frontiers in Environmental Science*, 10. 10.3389/fenvs.2022.791052.

Hasan, A.B., Reza, A.H.M.S., Kabir, S., *et al.* (2020). Accumulation and distribution of heavy metals in soil and food crops around the ship breaking area in southern Bangladesh and associated health risk assessment. *SN Applied Sciences*, 2, 155. https://doi.org/10.1007/s42452-019-1933-y.

Hasanein, P. and Emamjomeh, A. (2019). Beneficial effects of natural compounds on heavy metal-induced hepatotoxicity. In: Watson, R.R. and Preedy, V.R. (Eds.), *Dietary Interventions in Liver Disease. Foods, Nutrients and Dietary Supplements*, pp. 345–355. https://doi.org/10.1016/B978-0-12-814466-4.00028-8.

Hegazy, A.M.S. and Fouad, U.A. (2014). Evaluation of lead hepatotoxicity; histological, histochemical and ultrastructural study. *Forensic Medicine and Anatomy Research*, 2, 70–79. https://doi.org/10.4236/fmar.2014.23013.

Heghedűș-Mîndru, G., Negrea, P., Trașcă, T.I., Ștef, D.S., Cocan, I., and Heghedűș-Mîndru, R.C. (2023). Food intake of macro and trace elements from different fresh vegetables taken from Timisoara Market, Romania—Chemometric analysis of the results. *Foods*, 12, 749. https://doi.org/10.3390/foods12040749.

Horowitz, Y., Greenberg, D., Ling, G., and Lifshitz, M. (2002). Acrodynia: A case report of two siblings. *Archives of Disease in Childhood*, 86, 453. https://doi.org/10.1136/adc.86.6.453.

Hughes, M.F. (2002). Arsenic toxicity and potential mechanisms of action. *Toxicological Letters*, 133, 1–16.

Ihedioha, J.N., Ujam, O.T., Nwuche, C.O., Ekere, N.R., and Chime, C.C. (2016). Assessment of heavy metal contamination of rice grains (*Oryza sativa*) and soil from Ada field, Enugu, Nigeria: Estimating the human health risk. *Human and Ecological Risk Assessment: An International Journal*. https://doi.org/10.1080/10807039.2016.1217390.

Imeri, R., Kullaj, E., Duhani, E., and Millaku, L. (2019). Concentrations of heavy metals in apple fruits around the industrial area of Mitrovica, Kosovo. *The Iraqi Journal of Agricultural Sciences*, 50, 256–266.

International Agency for Research on Cancer (IARC). (1993). *Monographs—Cadmium*. Lyon, France.

Jaishankar, M., Tseten, T., Anbalagan, N., Mathew, B.B., and Beeregowda, K.N. (2014). Toxicity, mechanism and health effects of some heavy metals. *Interdisciplinary Toxicology*, 7, 60–72. https://doi.org/10.2478/intox-2014-0009.

Järup, L. (2003). Hazards of heavy metal contamination. *British Medical Bulletin*, 68, 167–182.

Junhe, L.U., Yang, X., Meng, X., Wang, G., Lin, Y., Wang, J., and Zhao, F. (2017). Predicting cadmium safety thresholds in soils based on cadmium uptake by Chinese cabbage. *Pedosphere*, 27, 475–481.

Kanagamani, K., Geethamani, P., and Narmatha, M. (2021). *Hazardous Waste Management*. IntechOpen. doi: 10.5772/intechopen.94080.

Kasam, R.S., Iresha, F.M., Wacano, D., Fauziah, I.F., and Amrullah, M.A. (2018). Evaluation of heavy metal exposure to soil and paddy plant around the closed municipal solid waste landfill: Case study at Gunung Tugel Landfill, Banyumas-Central Java. *IOP Conference Series: Materials Science and Engineering*, 299, 9. doi: 10.1088/1757-899X/299/1/012012.

Khan, M.U., Malik, R.N., and Muhammad, S. (2013). Human health risk from heavy metal via food crops consumption with wastewater irrigation practices in Pakistan. *Chemosphere*, 93, 2230–2238.

Khan, S., Cao, Q., Zheng, Y.M., Huang, Y.Z., and Zhu, Y.G. (2008). Health risks of heavy metals in contaminated soils and food crops irrigated with wastewater in Beijing, China. *Environmental Pollution*, 152, 686–692.

Kim, R.Y., Yoon, J.K., Kim, T.S., *et al.* (2015). Bioavailability of heavy metals in soils: Definitions and practical implementation—A critical review. *Environmental Geochemistry and Health*, 37, 1041–1061. https://doi.org/10.1007/s10653-015-9695-y.

Kim, Y.J. and Kim, J.M. (2015). Arsenic toxicity in male reproduction and development. *Development and Reproduction*, 19, 167–180. https://doi.org/10.12717/dr.2015.19.4.167.

Lentini, P., Zanoli, L., Granata, A., Signorelli, S.S., Castellino, P., and Dell'Aquila, R. (2017). Kidney and heavy metals—The role of environmental exposure

(Review). *Molecular Medicine Reports*, 15, 3413–3419. https://doi.org/10.3892/mmr.2017.6389.

Liu, L., Hu, L.L., Tang, J.J., Li, Y.F., Zhang, Q., and Chen, X. (2012). Food safety assessment of planting patterns of four vegetable type crops grown in soil contaminated by electronic waste activities. *Journal of Environmental Management*, 93, 22–30.

Ma, Y., Ran, D., Shi, X., Zhao, H., and Liu, Z. (2021). Cadmium toxicity: A role in bone cell function and teeth development. *Science of the Total Environment*, 769, 144646.

Matsunami, J. (2010). Hundred years of cadmium poisoning: Recollection and prospects. Katsura Shobo, Toyama. In: Kaji, M. (2012). Role of experts and public participation in pollution control: The case of Itai-itai disease in Japan. *Ethics in Science and Environmental Politics*, 12, 99–111.

Medvedeva, Y., Kucher, A., Lipsa, J., and Hełdak, M. (2021). Human health risk assessment on the consumption of apples growing in urbanized areas: Case of Kharkiv, Ukraine. *International Journal of Environmental Research and Public Health*, 18, 1504. doi: 10.3390/ijerph18041504.

Melgar, M.J., Alonso, J., and Garcia, M.A. (2009). Mercury in edible mushrooms and underlying soil: Bioconcentration factors and toxicological risk. *Science of the Total Environment*, 407, 5328–5334.

Messner, B., Knoflach, M., Seubert, A., Ritsch, A., Pfaller, K., Henderson, B., Shen, Y.H., Zeller, I., Willeit, J., Laufer, G., Wick, G., Kiechl, S., and Bernhard, D. (2009). Cadmium is a novel and independent risk factor for early atherosclerosis mechanisms and in vivo relevance. *Arteriosclerosis, Thrombosis, and Vascular Biology*, 29, 1392–1398. https://doi.org/10.1161/ATVBAHA.109.190082.

Milton, A., Hussain, S., Akter, S., Rahman, M., Mouly, T., and Mitchell, K. (2017). A review of the effects of chronic arsenic exposure on adverse pregnancy outcomes. *International Journal of Environmental Research and Public Health*, 14, 556. https://doi.org/10.3390/ijerph14060556.

Mirecki, N., Agic, R., Šunić, L., Milenkovic, L., and Ilic, Z. (2015). Transfer factor as indicator of heavy metals content in plants. *Fresenius Environmental Bulletin*, 24, 4212–4219.

Mirsal, I.A. (2008). *Soil Pollution: Origin, Monitoring & Remediation*. Springer, 2nd edition, p. 327.

Mishra, K.P. (2009). Lead exposure and its impact on immune system: A review. *Toxicology in Vitro*, 23, 969–972. https://doi.org/10.1016/j.tiv.2009.06.014.

Mitra, S., Chakraborty, A., Tareq, A., Emran, T., Nainu, F., Khusro, A., Idris, A., Khandaker, M., Osman, H., Alhumaydhi, F., and Simal-Gandara, J. (2022). Impact of heavy metals on the environment and human health: Novel therapeutic insights to counter the toxicity. *Journal of King Saud University—Science*, 34, 101865. https://doi.org/10.1016/j.jksus.2022.101865.

Mongi, R. and Chove, L. (2020). Heavy metal contamination in cocoyam crops and soils in countries around the Lake Victoria Basin (Tanzania, Uganda and Kenya). *Tanzania Journal of Agricultural Sciences*, 19, 148–160.

Moszynski, P. (2010). Lead poisoning in Nigeria causes "unprecedented" emergency. *BMJ*, 341, c4031. https://doi.org/10.1136/bmj.c4031.

Munir, N., Jahangeer, M., Bouyahya, A., El Omari, N., Ghchime, R., Balahbib, A., Aboulaghras, S., Mahmood, Z., Akram, M., and Ali Shah, S.M. (2022). Heavy metal contamination of natural foods is a serious health issue: A review. *Sustainability*, 14, 161. https://doi.org/10.3390/su14010161.

Navas-Acien, A., Selvin, E., Sharrett, A.R., Calderon-Aranda, E., Silbergeld, E., and Guallar, E. (2004). Lead, cadmium, smoking, and increased risk of peripheral arterial disease. *Circulation*, 109, 3196–3201. https://doi.org/10.1161/01.CIR.0000130848.18636.B2.

Nyiramigisha, P. and Komariah, S. (2021). Harmful impacts of heavy metal contamination in the soil and crops grown around dumpsites. *Reviews in Agricultural Science*, 9, 271–282. https://dx.doi.org/10.7831/ras.9.0_271.

Obiora, S.C., Chukwu, A., and Davies, T.C. (2016). Heavy metals and health risk assessment of arable soils and food crops around Pb-Zn mining localities in Enyigba, southeastern Nigeria. *Journal of African Earth Sciences*. https://doi.org/10.1016/j.jafrearsci.2015.12.025.

Olaniran, A.O., Balgobind, A., and Pillay, B. (2013). Bioavailability of heavy metals in soil: Impact on microbial biodegradation of organic compounds and possible improvement strategies. *International Journal of Molecular Sciences*, 14, 10197–10228. https://doi.org/10.3390/ijms140510197.

Orisakwe, O., Mbagwu, H., Ajaezi, G., Edet, U., and Uwana, P. (2015). Heavy metals in seafood and farm produce from Uyo, Nigeria: Levels and health implications. *Sultan Qaboos University Medical Journal*, 15, 275–282.

Orloff, K. and Falk, H. (2003). An international perspective on hazardous waste practices. *International Journal of Hygiene and Environmental Health*, 206, 291–302. doi: 10.1078/1438-4639-00225.

Osae, R., Nukpezah, D., Darko, D.A., Koranteng, S.S., and Mensah, A. (2023). Accumulation of heavy metals and human health risk assessment of vegetable consumption from a farm within the Korle lagoon catchment. *Heliyon*, 9, e16005. https://doi.org/10.1016/j.heliyon.2023.e16005.

Owonikoko, W.M. and Alimba, C.G. (2024a). Systematic literature review of heavy metal contamination of the Nigerian environment from e-waste management: Associated health and carcinogenic risk assessment. *Toxicology*, 505, 153811. https://doi.org/10.1016/j.tox.2024.153811.

Owonikoko, W.M. and Alimba, C.G. (2024b). Heavy metal contamination of the Nigerian environment from e-waste management: A systematic review of exposure pathway and attendant pathophysiological implications. *Toxicology*, 509, 153966. https://doi.org/10.1016/j.tox.2024.153966.

Patrick, L. (2002). Mercury toxicity and antioxidants: Part 1: Role of glutathione and alpha-lipoic acid in the treatment of mercury toxicity. *Alternative Medicine Review*, 7, 456–471.

Peters, J.L., Perlstein, T.S., Perry, M.J., McNeely, E., and Weuve, J. (2010). Cadmium exposure in association with history of stroke and heart failure. *Environmental Research*, 110, 199–206. https://doi.org/10.1016/j.envres.2009.12.004.

Pierce, B.L., Kibriya, M.G., Tong, L., Jasmine, F., Argos, M., Roy, S., Paul-Brutus, R., Rahaman, R., Rakibuz-Zaman, M., Parvez, F., Ahmed, A., Quasem, I., Hore, S.K., Alam, S., Islam, T., Slavkovich, V., Gamble, M.V., Yunus, M.D., Rahman, M., Baron, J.A., Graziano, J.H., Ahsan, H., and McCarthy, M.I. (2012). Genome-wide association study identifies chromosome 10q24.32 variants associated with arsenic metabolism and toxicity phenotypes in Bangladesh. *PLoS Genetics*, 8, e1002522. https://doi.org/10.1371/journal.pgen.1002522.

Polprasert, C. and Liyanage, L.R.J. (1996). Hazardous waste generation and processing. *Resources, Conservation and Recycling*, 16, 213–226.

Pouillot, R., Santillana, F.S., and Van Doren, J.M. (2022). Modeling the risk of low bone mass and osteoporosis as a function of urinary cadmium in U.S adults aged 50–79 years. *Environmental Research*, 212, 113315.

Prüss-Üstin, A. and Corvalán, C. (2006). *Preventing Disease Through Healthy Environments. TOWARDS an Estimate of the Environmental Burden of Disease*. World Health Organization, Geneva.

Rahimzadeh, R.M., Kazemi, S., and Moghadamnia, A.A. (2017). Cadmium toxicity and treatment: An update. *Caspian Journal of Internal Medicine*, 8, 135–145. https://doi.org/10.22088/cjim.8.3.135.

Rai, P.K., Lee, S.S., Zhang, M., Tsang, Y.F., and Kim, K.H. (2019). Heavy metals in food crops: Health risks, fate, mechanisms, and management. *Environment International*, 125, 365–385. https://doi.org/10.1016/j.envint.2019.01.067.

Reczek, C.R. and Chandel, N.S. (2017). The two faces of reactive oxygen species in cancer. *Annual Review of Cancer Biology*, 1, 79–98. https://doi.org/10.1146/annurevcancerbio-041916-065808.

Rosenfeld, P.E. and Feng, L.G.H. (2011). *Risks of Hazardous Wastes*. Elsevier Inc., p. 472.

Rosli, N.A., Aziz, H.A., Pueh, L.L.L., Othman, I.B., Zawawi, M.H., and Hung, Y.T. (2024). Management of various sources of hazardous waste. In: Wang, L.K., Sung Wang, M.H., and Hung, Y.T. (Eds.), *Waste Treatment in the Biotechnology, Agricultural and Food Industries. Handbook of Environmental Engineering*, 27. Springer, Cham. https://doi.org/10.1007/978-3-031-44768-6_2.

Sanga, V.F. and Pius, C.F. (2024). Heavy metal contamination in soil and food crops and associated human health risks in the vicinity of Iringa Municipal dumpsite, Tanzania. *Discover Environment*, 2, 104. https://doi.org/10.1007/s44274-024-00137-y.

Sedman, R.M. and Esparza, J.R. (1991). Evaluation of volatile organic emissions from hazardous waste incinerators. *Environmental Health Perspectives*, 94, 169–180. https://doi.org/10.1289/ehp.94-1567964.

Sharafi, K., Omer, A., Mansouri, B., Massahi, T., Soleimani, H., Moradi, M., Parnoon, K., and Ebrahimzadeh, G. (2024). Transfer of heavy metals from soil to vegetables: A comparative assessment of different irrigation water sources. *Heliyon*, 10, e32575. 10.1016/j.heliyon.2024.e32575.

Shukla, S., Ganguly, R., and Hussain, C.M. (2020). The handbook of environmental remediation: Classic and modern techniques. In: Hussain, C.M. (ed). *The Handbook of Environmental Remediation*. The Royal Society of Chemistry, pp. 24–52.

Singh, N., Kumar, D., and Sahu, A. (2007). Arsenic in the environment: Effects on human health and possible prevention. *Journal of Environmental Biology*, 28, 359–365.

Smical, A.I., Hotea, V., Oros, V., Juhasz, J., and Pop, E. (2008). Studies on transfer and bioaccumulation of heavy metals from soil into lettuce. *Environmental Engineering and Management Journal*, 7, 609–615.

Tchounwou, P.B., Patlolla, A.K., and Centeno, J.A. (2003). Carcinogenic and systemic health effects associated with arsenic exposure—A critical review. *Toxicologic Pathology*, 31, 575–588.

Tchounwou, P.B., Yedjou, C.G., Patlolla, A.K., and Sutton, D.J. (2012). Heavy metal toxicity and the environment. *Experientia Supplementum*, 101, 133–164. https://doi.org/10.1007/978-3-7643-8340-4_6.

Tellez-Plaza, M., Navas-Acien, A., Crainiceanu, C.M., and Guallar, E. (2008). Cadmium exposure and hypertension in the 1999–2004 National Health and Nutrition Examination Survey (NHANES). *Environmental Health Perspective*, 116, 51–56. https://doi.org/10.1289/ehp.10764.

Tirima, S., Bartrem, C., von Lindern, I., von Braun, M., Lind, D., Anka, S.M., and Abdullahi, A. (2018). Food contamination as a pathway for lead exposure in children during the 2010–2013 lead poisoning epidemic in Zamfara, Nigeria. *Journal of Environmental Science*, 67, 260–272. https://doi.org/10.1016/j.jes.2017.09.007.

U.S. EPA. (2012). A citizen's guide to incineration. https://www.epa.gov/sites/default/files/201504/documents/a_citizens_guide_to_incineration.pdf. Accessed 28 October 2024.

Udofia, U.U., Udiba, U.U., Akpan, E.R., and Antai, E.E. (2020). Potential human health risk assessment of heavy metals intake through consumption of fluted pumpkin (*Telfairia occidentalis*) purchased from major markets in Calabar Metropolis, Nigeria. *Issues in Biological Sciences and Pharmaceutical Research*, 8, 85–97.

US EPA. (2005). Introduction to hazardous waste identification, 40 CFR Part 261, Solid Waste and Emergency Response, 5305W, US Environmental

Protection Agency, Washington, DC, EPA/530/K/05-012. https://www.epa.gov/sites/production/files/2015-09/documents/hwid05.pdf. Accessed October 2024.

Uzakov, Z. and Uzakov, Z. (2023). Accumulation of heavy metals in the soil and agricultural crops. *BIO Web of Conferences*, 71, 01065. https://doi.org/10.1051/bioconf/20237101065.

Valko, M., Morris, H., and Cronin, M.T.D. (2005). Metals, toxicity, and oxidative stress. *Current Medicinal Chemistry*, 12, 1161–1208.

Vanisree, C.R., Singh, S., Singh, M., Jadhav, P.B., Kumar, E., Verma, R., Awasthi, K., and Nagar, V. (2022). *Heavy Metal Contamination of Food Crops: Transportation Via Food Chain, Human Consumption, Toxicity and Management Strategies*. IntechOpen. doi: 10.5772/intechopen.101938.

Visvanathan, C. (1996). Hazardous waste disposal. *Resources, Conservation and Recycling*, 16, 201–212.

World Health Organization (WHO). (2016). Waste and human health: Evidence and needs. WHO Meeting Report: 5–6 November 2015. Bonn, Germany. WHO Regional Office for Europe, Copenhagen.

Wu, D., Yanjun, L., Lingyi, M., Guilan, L., Yazhuo, C., and Ziliang, Y. (2024). Distribution, occurrence, and environmental risks of heavy metals in hazardous waste: A regional study in Beijing, China. *Circular Economy*, 3, 100099.

Yan, A., Wang, Y., Tan, S.N., Mohd, Y.M.L., Ghosh, S., and Chen, Z. (2020). Phytoremediation: A promising approach for revegetation of heavy metal-polluted land. *Frontiers in Plant Science*, 11. doi: 10.3389/fpls.2020.00359.

Yang, Y., Chang, A.C., Wang, M., Chen, W., and Peng, C. (2018). Assessing cadmium exposure risks of vegetables with plant uptake factor and soil property. *Environmental Pollution*, 238, 263–269.

Zangina, A.S. and Ali, A.F. (2021). An overview of hazardous waste management in Nigeria. *Dutse Journal of Pure and Applied Sciences (DUJOPAS)*, 7, 214–221.

https://doi.org/10.1142/9789819822249_0005

Chapter 5

Health Risks Linked to Hazardous Waste-Related Metal Pollution in Foods

Julekha Khatun* **and Debasis Dhak*,†,‡**

**Nanomaterials Research Lab, Department of Chemistry, Sidho-Kanho-Birsha University, Purulia 723104, India*

†debasisdhak@yahoo.co.in

‡debasis.chem@skbu.ac.in

Abstract

Agriculture and industry play a vital role in this era of increasing population growth by giving millions of people access to food and jobs. However, the requirement for quick and high-quality plant production is becoming increasingly necessary to fulfill the world's ever-increasing food needs. One such action that humans have long been used to performing, for this reason, is fertilization. However, the excessive use of chemical fertilizers and industrial wastes from different industries is having a detrimental effect on the ecosystem and general public health. All heavy metals (HMs) are dangerous, but because of their abundance in the environment, certain of them—such as arsenic (As), lead (Pb), and cadmium (Cd)—are particularly important. They enter the food chain through various pathways, including soil to plants to herbivores to carnivores to humans, water to aquatic organisms to fish to humans, and air to plants to herbivores to carnivores to humans. Consuming contaminated food can lead to neurological damage (cognitive impairment and memory loss), cancer (carcinogenic effects), kidney damage (nephrotoxicity), cardiovascular disease, reproductive problems (infertility and birth defects), and respiratory issues (asthma and chronic obstructive pulmonary disease). This chapter gives detailed information on the health risks linked to hazardous waste-related metal pollution in foods.

Keywords: Hazardous wastes, fertilization, industrialization, toxic heavy metals (HMs), HM sources, toxic effects.

1. Introduction

The alarming rise of industrialization has been accompanied by unprecedented economic growth, technological advancements, and improved living standards. However, this progress comes at a steep environmental and health cost. The unchecked discharge of hazardous waste, particularly HMs, has become a pressing global concern. These toxic substances, omnipresent in modern industrial processes, contaminate our soil, water, air, and food, posing existential threats to human health, ecosystems, and biodiversity. For their food-making process, photosynthesis, the majority of plants obtain water and nutrients from the soil. We eat plants directly or consume animals that eat plants. According to the food chain, the whole of mankind is therefore totally dependent on the soil. Fertilization becomes one of the most vital methods of obtaining food as the world's population grows and the need to fulfill the enormous demand for it grows. However, it has been shown that overfertilization of permanent lands degrades the quality of the soil (Savci, 2012). On the other side, using pesticides to eradicate pests is crucial to safeguarding farm output. India has been ranked 12th in the world and at the top of the Asian pesticide production ranking (Sharma and Singhvi, 2017). Overuse of fertilizers lowers the pH of the soil, which increases the availability of naturally existing HMs (Khan *et al.*, 2017).

HMs, a diverse group of elements characterized by their high density and toxicity, are ubiquitous in industrial waste streams. Pb, Hg, As, Cd, and Cr are among the most prevalent and hazardous HMs, with far-reaching consequences for environmental and human health (Sonone *et al.*, 2021). Their persistence in the environment, coupled with their ability to bioaccumulate and biomagnify, has devastating impacts on ecosystems, food chains, and human populations. While high concentrations of other metals, such as cadmium (Cd), chromium (Cr), manganese (Mn), and lead (Pb), are thought to be extremely toxic for humans and aquatic life, certain HMs, such as copper (Cu), cobalt (Co), and zinc (Zn), are fundamentally necessary for normal bodily growth and functions of living organisms (Ouyang *et al.*, 2002). Normal physiological processes require a certain amount of Cr, but large quantities can be hazardous and result in genotoxic carcinogens as well as liver and renal issues. As a metal component of vitamin B12, Co is likewise necessary for regular bodily processes, much like Cr (Strachan, 2010). However, polycythaemia, aberrant thyroid arteries, excessive red blood cell (RBC) production, and issues with the right coronary artery might result from consuming contaminated food and water that contains high levels of Co (Muhammad *et al.*, 2011). In general, mental illnesses such as Alzheimer's and manganism can be brought on by elevated levels of Mn and Cu in drinking water (Dieter *et al.*, 2005). The cognitive abilities of 10-year-old children are similarly impacted by elevated Mn pollution in drinking water, according to Wasserman *et al.* (Wasserman *et al.*, 2006). In a similar vein, consuming Ni-sulfate and Ni-chloride can result in serious health issues, including cardiac arrest (Knight *et al.*, 1997). In addition, lead is a very toxic and carcinogenic metal that can lead to gliomas, lung cancer, stomach cancer, nerve damage, kidney damage, headache, irritability, and abdominal discomfort, among other chronic health hazards (Muhammad *et al.*, 2011). Children are particularly prone to Pb poisoning, and prolonged exposure to high amounts of Pb can result in anaemia as well as serious health complications, such as behavioral abnormalities, memory loss, and diminished comprehension (Muhammad *et al.*, 2011). Similar to other HMs, getting enough zinc is crucial for regular bodily processes. Poor wound healing, decreased respiratory muscle function, immunological dysfunction, anorexia, diarrhea, hair loss, dermatitis (acrodermatitis enteropathica), and depression might result from its insufficiency. On the other hand, sideroblastic anaemia is brought on by zinc poisoning (Strachan, 2010). In animals and plants, exposure to cadmium can have both immediate and long-term health impacts (Barbee and Prince, 1999). Chronic consequences include itai-itai (ouch-ouch) disorders, renal damage, and bone damage (Muhammad *et al.*, 2011). Cd may induce cancer in humans, according to experimental studies in both people and animals (Muhammad *et al.*, 2011). The nature and severity of toxicity are influenced by the type of heavy metal involved, the amount of exposure, the chemical and valence states (inorganic versus organic), the manner of exposure (acute versus chronic), and the subject's age. Together, soil, water, and air make up nature. Therefore, one ingredient's contamination can contaminate the others (Harrison and Alloway, 2007). Additionally contributing to soil contamination include leather products, food scraps, plastics, radioactive waste, and chemicals, including zinc, aluminium, cyanides, cadmium, alkalies, and acids, found in industrial wastes and organic waste (Harrison and Alloway, 2007). The entire population of humans is thus entirely reliant on the soil and water, according to the food chain (Khatun *et al*., 2022). Thus, their contamination also contaminates the food and penetrates in to the human body leading to severe disease also leading to death (Khatun *et al*., 2022). The World Health Organization (WHO) and the United Nations Environment Programme (UNEP) have identified HM pollution as a significant risk factor for various diseases, including cancer, neurological disorders, reproductive problems, and respiratory diseases. Moreover, the economic costs of heavy metal pollution are substantial, with estimated annual losses in the trillions of dollars.

This chapter mainly focuses on the health risks in the aquatic environment and also in human health linked to hazardous waste-related metal pollution in foods.

2. Sources of HMs

Elements with a higher atomic mass and density that can have an impact on both individuals as well as the environment are known as heavy metals (HMs) (Zaynab *et al*., 2022). Since heavy metals are not biodegradable, they can be dangerous even at low amounts (Brodin *et al*., 2017; Ferrey *et al*., 2018). Due to absorption, inhalation, mistreatment, accidents, and other factors, heavy metals (HMs) accumulate up in soil, human, and animal tissues. Due to normal biogeochemical cycles, metals have existed on Earth since the beginning. HMs were found in the soil as a result of the underlying weathering mechanism. The soil in the Mendip area of Great Britain is rich in cadmium, lead, and zinc due to mineralized veins and metal deposits in high concentrations in the bedrock. Bedrock weathering with a slightly elevated concentration of heavy metals (HMs) can lead to metal enrichment during soil formation.

Anthropogenic and human factors are the primary contributors to the elevated environmental toxicity associated with heavy metals. Windblown soil debris, biogenic processes, forest fires, volcanic eruptions, and

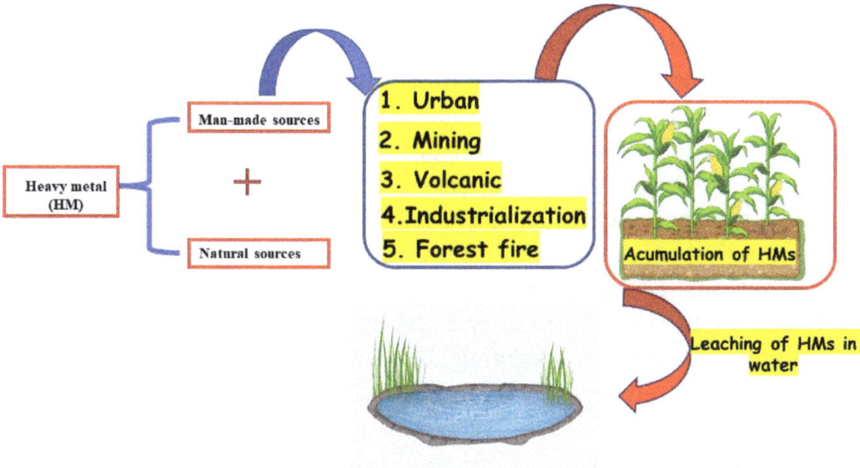

Fig. 1. Accumulation of HMs in soil and water from their sources.

sea salt are examples of natural sources of heavy metals (HMs) (Muhammad *et al*., 2011; Blaser *et al*., 2000). Mining activities, the use of pesticides, fertilizers, and herbicides, as well as the irrigation of agricultural fields with sewage and industrial water, are anthropogenic sources of HMs pollution (Srivastava *et al*., 2018). One of the main sources of heavy metal contamination in our food is trace amounts of HMs in fertilizers. Anthropogenic practices that cause HMs contamination include improper industrial waste management, traffic pollution, lead (Pb) usage asaerosol cans, fuel antiknock, metallurgy and smelting, sewage discharge, and construction materials (Zaynab *et al*., 2022; Srivastava *et al*., 2016) Accumulation of HMs in soil and water from their sources is shown in Fig. 1.

3. Toxic effects of HMs

Contamination by heavy metals is becoming a worldwide problem. The gills, the body surface, and the digestive tract are the three ways that heavy metals can enter fish (Dane and Şişman, 2020). According to Krieger *et al*. (2020), fish juveniles and larvae grow quite quickly, and their development in terms of body length and mass is directly correlated with an appropriate temperature and an adequate quantity of food. On the other side, poisonous diet containing heavy metals hinders fish growth. Growth inhibition is among the most visible indicators of metal poisoning in fish. Therefore, by changing different enzymes and metabolites, HM concentrations in tissues lead to a range of metabolic, physiological, and histological alterations in fish and other freshwater species (Mehmood *et al.*, 2019). Depending on a number of variables, including developmental agents, psychological agents, and fish longevity, different species have different eating mechanisms. Fish living in contaminated environments have tissues that collect heavy metals (HMs) (Kumar *et al.*, 2020; Topal and Onac, 2020). The choice of body organs for HMs deposition depends on a number of criteria, including metal intensity, expression duration, metal absorption, environmental factors (temperature, pH, hardness, and salinity), and intrinsic agents, such as fish age and eating habits. The kidneys, gills, and liver are where the majority of metals build up (Kucukosmanoglu and Filazi, 2020; Squadrone *et al.*, 2020). When zinc accumulates up in fish gills, it disrupts the oxygen flow to tissues, resulting in hypoxia and death. However, HMs may be released into the water column and mobilized if the pH of the water drops, putting marine life—including insects and crustaceans—at risk (Bonsignore *et al.*, 2018). The benthic species are killed by these poisonous sediments, which also decreases the amount of food available to the enormous monster. HMs present in food and the environment are essential for good health at low concentrations, but they can be hazardous or dangerous at high concentrations. Their toxicity can harm the brain, liver, lungs, kidneys, blood, and other essential organs in addition to depleting energy. Degenerative physical, tissue, and neurological processes that mimic illnesses including multiple sclerosis, Parkinson's, Alzheimer's, and muscular dystrophy ultimately arise from prolonged exposure. Reduced appetite, headaches, high blood pressure, nausea, kidney problems, exhaustion, sleeplessness, arthritis, hallucinations, and dizziness can all be symptoms of acute lead (Pb) exposure. Acrodynia or pink illness is caused by mercury poisoning. According to Guzzi *et al.* (2021), elevated mercury exposure may alter the structure of the brain and result in visual or auditory impairments, tremors, cognitive decline, irritability, and shyness. Shorter periods of exposure to greater concentrations of metallic mercury vapors may cause skin rashes, elevated blood pressure, nausea, vomiting, diarrhea, and lung damage. Signs and symptoms of organic mercury poisoning include hair loss, headaches, tremors, depression, and memory issues. It may be challenging to identify some situations since these symptoms and indicators are commonly linked to other illnesses (Atti *et al.*, 2020).

4. Bioavailability of HMs in Food Webs

Fish bioaccumulate hazardous substances as a result of heavy metal pollution of rivers, lakes, and streams. A variety of pathways, such as food intake and the assimilation of sediment particles, may allow HMs to enter fish (Liu *et al.*, 2020). Fish and other aquatic animals rely on a variety of invertebrates as food sources, and they offer a useful pathway for the absorption of lead, copper, zinc, and cadmium (Corrias *et al.*, 2020; Jardine *et al.*, 2020). Another way that these harmful substances might be absorbed is by immediate water absorption (Maurya *et al.*, 2019). According to Luoma and Rainbow (2008) and Suter and Suter (2009), sediment offers a third potential pathway and is the main reservoir of trace elements in maritime environments. The Bioaccumulation Factor (BAF), which measures the rate of element concentration between fish and abiotic (water and sediment) habitats, was commonly used to evaluate the pollution condition of water bodies (Mortuza and Al-Misned, 2015; Ziyaadini *et al.*, 2017). The food chain is contaminated by significant and trace elements that are transferred from abiotic to living species in this setting and accumulate in biota (Ali and Khan, 2018). In food chains, organisms at higher trophic levels are more susceptible to biomagnification. Higher concentrations of trace elements in species with higher trophic levels can put these creatures or people in danger due to bioamplification. Smaller and bigger animals eventually consume the methyl, dimethyl, and ethyl mercury produced by the activities of water-living microbes (Bernhoft, 2012). Aquatic bodies have been shown to have over 20 different organic and inorganic chemical compounds that include arsenic (As). Numerous elements, including bacteria, phytoplankton, salinity, temperature, and redox conditions, affect the shape of chemical compounds found in water (Zhang *et al.*, 2013). One source of arsenic can be changed into another by marine microbes. Invertebrates that deposit arsenic become part of the marine food chain (Casado-Martinez *et al.*, 2010). Higher species rely on these invertebrates as a significant source of aquatic food, and when fish eat these invertebrates, they absorb arsenic particles. The results of the study indicate that fish are the main source of human exposure to arsenic (Juma *et al.*, 2002). According to Nair *et al.* (2006), lead is toxic to marine life, and fish, which are at the top of the aquatic food chain, accumulate lead in their livers and gills at a rapid rate. Fish is commonly consumed by people as part of their daily diet. HMs build up in the body because lead is absorbed, moved into the bloodstream, and then integrated into tissues (Nair *et al.*, 2006; Castro-González and Méndez-Armenta, 2008).

5. Bioaccumulation of HMs in Seafood and Effect on Aquatic Environment

Significant ecological, environmental, and societal ramifications result from the bioaccumulation of heavy metals in freshwater fish (Ali *et al.*, 2017). Higher quantities of metals are accumulated by animals when they are present in the environment, resulting in the biomagnification of metals in the trophic web, which negatively affects aquatic ecology since it depends on them directly or indirectly in a number of ways (Suter and Suter, 2009). Freshwater fishes and other aquatic life have decreased in Pakistan's Indus River as a result of increased pollution (Al-Ghanim *et al.*, 2016). Environmental factors such as HMs bioavailability and concentration in water columns, water physicochemical properties, and other climatic factors, as well as fish characteristics such as size, sex, reproductive cycle, feeding habits, and swimming patterns, all play a crucial role in the bioaccumulation of HMs (Ali *et al.*, 2019; Moiseenko and Gashkina, 2020). Depending on the structure and function of the tissue, the degree of HMs deposition in different fish organs usually varies. Compared to other tissues such as the skin and muscles, metabolically active tissues such as the kidneys, liver, or gills often acquire more heavy metals (HMs). It has been determined that HMs such as nickel (Ni) aggregate and are disposed of in fish gills (El-Moselhy *et al.*, 2014; Mansouri *et al.*, 2012). Fish muscles bioaccumulate HMs in a species-specific way (Chakraborty *et al.*, 2016). Indian anchovies (Stolephorusindus), which were gathered from the Arabian Gulf in the United Arab Emirates, were found to have toxic substances. Significant concentrations of zinc were observed mostly in the muscles, while the liver and muscles had significant concentrations of Cd, Cu, and Cr (Alizada *et al.*, 2020). The gills, eyes, stomach, liver, gonads, and muscles of three fish species—Labeorohita, Pangasius hypophthalmus, and Katsukiwonus pelamis—that were collected from Visakhapatnam were examined. Co, Hg, and Pb were below the detectable limit (BDL), but the metal concentrations in small and big *L. rohita* and *P. hypophthalmus* were in the order Fe > Zn > Cu. Fe > Zn > Cu > Cd were the HMs concentrations in K. pelamis, while Pb, Hg, and Co were BDL (Pragnya *et al.*, 2020). Bivalves are vital to the human diet and play a significant role in the biogeochemical cycle as filter feeders of the aquatic environment. By consuming saltwater, they can collect heavy metals (HMs), which they can then use as food for other marine organisms at higher trophic levels (Kodama *et al.*, 2012; Yuan *et al.*, 2020). Bivalve molluscs have the ability to absorb pollutants from the water column, suspended particulate matter, sediments, and even food sources. Bivalves' rate of metal bioaccumulation is influenced by both biotic (such as organisms, age, sex, weight, gametogenesis, and physiological state) and abiotic (such as chemical species, pH, salinity, temperature, filtration rate, and the presence of environmental toxins) variables. The ability of bivalves to bioaccumulate heavy metals was high (Yuan *et al.*, 2020). Human health suffers when edible bivalves are consumed. The Ocypodid family of crabs are depositary feeders that are essential prey and a staple diet for many aquatic birds and animals. They play a major role in organic mineralization, sediment purification, and particle size reduction (Gouws and Stewart, 2001). According to studies, aquatic creatures in Tuticorin's coastal zones experienced the bioaccumulation of heavy metals, such as Zn, Cu, Cd, and Pb. Crabs in polluted environments are important bioaccumulators of heavy metals, claim Yogeshwaran *et al.* (2020). The purpose of the study was to investigate the Fe, Cu, Zn, Cr, Ni, Co, Pb, and Cd accumulation and biomonitoring potential of HMs in Macrophthalmus depressus from Karachi, Pakistan (Saher and Siddiqui, 2019).

5.1. *Effect on gills and respiratory damage*

There are three possible entry points for heavy metals into fish bodies: the gills, the digestive tract, and the body surface. One prominent location for the major intake of metals from the water is the gills (Sonone *et al.*, 2021). Fish have gills that collect zinc. It suggests that tissue respiration will get discouraged and eventually develop into hypoxia, which would result in mortality. Alterations in heart and ventilator architecture are also triggered by zinc exposure (Olaifa *et al.*, 2010). The essential toxicant zinc damages fish gill tissue, impairs ion and acid-base balance, and results in hypoxia (Vosylienė and Jankaitė, 2006; Senthil Murugan *et al.*, 2008). Numerous blood abnormalities, including eosinophilia, anaemia, lymphocytosis, and renal and bronchial diseases, are indicative of the harmful effects of chromium in some fish species. Higher concentrations of Cr harm the gills of fish swimming close to the site of Cr dispersion, but chromium is known to accumulate very little in fish bodies (Sonone *et al.*, 2021).

5.2. *Effect on blood and circulatory system*

In little amounts, nickel is necessary for the formation of red blood cells (RBCs), but in large amounts, it becomes undetectable poisonous. In fish, cadmium has been linked to an increase in blood pressure and heart disease (Govind and Madhuri, 2014). Research indicates that mercury causes damage to fish blood vessels, haemorrhages, and a decrease in the number of blood cells in the fish. After anaemia, eosinophilia, lymphocytosis, bronchial, and renal injuries, chromium affects the blood (Sonone *et al*., 2021).

5.3. *Genotoxic effect*

Proteins are coagulated by arsenic, which also forms bonds with coenzymes and inhibits the production of adenosine triphosphate (ATP) during respiration (Sankhla, 2019). In aggregates of all its oxidation states, it may induce cancer, and excessive levels of expression can be fatal (Sankhla, 2019). Numerous studies indicate that heavy metals may have genotoxic effects in addition to their poorly understood carcinogenic effects (Sonone *et al*., 2021). In addition to inhibiting DNA repair, arsenic composites cause chromosome abnormalities, sister-chromatid transfers, and the formation of micronuclei in cells (Sonone *et al*., 2021). DNA degradation by base-pair deletion, mutation, or radical oxygen assault on DNA has been linked to carcinogenic metals, such as cadmium, arsenic, nickel, and chromium (Landolph, 1994).

5.4. *Effect on reproductive health*

Mercury is a significant heavy metal for neurotoxicity. Although there are additional factors that contribute to neurotoxicity, mercury has a major impact on both people and fish. Monomethyl Hg causes brain damage, and its foetal and postnatal manifestations have resulted in congenital deformities, abortions, and developmental abnormalities in early fry. Japanese streams are contaminated by mercury. Significant neurotoxicity was shown by Minamata illness and mercury poisoning (by methyl mercury) (Govind and Madhuri, 2014). Higher levels of zinc, lead, cadmium, copper, mercury, and cadmium were found in fish gonads, according to a number of research conducted on fish species, including Oreochromis mossambicus, Noemacheilus barbatulus, Catostomus commersoni, Perca fluviatilis, and Oreochromis aureus. Fish fertility and embryonic

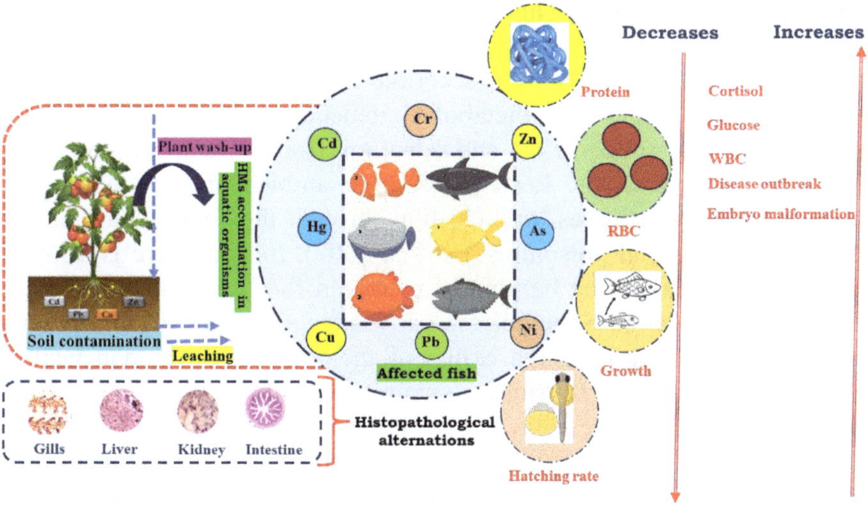

Fig. 2. Impact of HMs on aquatic health.

development were doubtfully impacted, and eggs and sperm became contaminated as a result (Sonone *et al.*, 2021). Consuming methylmercury reduced sperm motility significantly. Both fish metabolism and embryo development have been affected by heavy metals (Sonone *et al.*, 2021). A smaller number of hatched larvae was the consequence of anatomical and functional abnormalities throughout embryonic development. Lead and copper absorptions caused the death of many freshly born larvae shortly after hatching (Jezierska *et al.*, 2009). Fish with heavy metal sensitivity have abnormal differentiation and have their oestrogenic and androgenic secretions corrupted. Numerous contaminants, including pesticides, heavy metals, and waste from industry and agriculture, have histopathological effects on fish gonadal reproductive tissues (Sonone *et al.*, 2021). These problems might interfere with the germ cell's development and reduce the fish's ability to reproduce (Sonone *et al.*, 2021). The HM toxicity effects on aquatic animals are shown in Fig. 2.

6. Intake of Heavy Metal Through Food Chain

We are more susceptible to these heavy metals through our food, water, and air. The metabolism depends on a few of these heavy metals, including copper, zinc, and selenium. But they can cause toxicity at higher concentrations. Lead pipes, high ambient air concentrations close to emission sources, or ingestion through the food chain can all result in heavy metal poisoning. We are more susceptible to these heavy metals through our food, water, and air. The metabolism depends on a few of these heavy metals, including copper, zinc, and selenium. But they can cause toxicity at higher concentrations. Lead pipes, high ambient air concentrations close to emission sources, or ingestion through the food chain can all result in heavy metal poisoning (Duffus, 2002). Indeed, there is a lot of interest worldwide in the buildup of metals in food crops and how they affect human health. Nonetheless, knowledge of geophysical changes might aid us in assessing the extent of their impact on human health. Along with the origin of metallic contaminants, which has received little interpretation, difficulties may differ between nations (Rai *et al.*, 2019). It is known that heavy metals are toxicants that cause acute illnesses in aquatic animals. Heavy metal absorption in aquatic species through the food chain as shown in Fig. 3 can result in a number of clinical dysfunctions, including human hypertension, cramping, occasional fever, and kidney impairment. Being at the base of the food pyramid and acting as acceptable transfer media for humans, fish play an important role in the

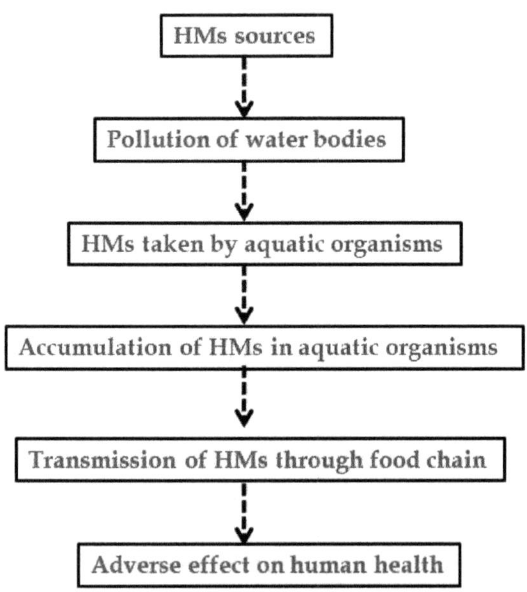

Fig. 3. Human intake of HMs through food chain.

biomagnification of metals (Sankhla, 2019; Oze *et al.*, 2009). Since they may cause oxidative degradation of biological macromolecules and have toxic and carcinogenic consequences, heavy metals can be extremely harmful to people.

7. Effect of HMs on Human Health

The environment naturally contains heavy metals. When present in trace amounts, metals are beneficial to humans. They are referred to be essential metals because they support biological functions, such as the creation of haemoglobin, the transfer of oxygen and electrons, cell metabolism, enzyme control, the generation of hormones and antioxidants, and human cell growth. On the other hand, increased levels of heavy metals have harmful consequences on people. Heavy metals can enter the human body primarily through ingestion (eating or drinking) and inhalation (breathing) through a variety of methods, including drinking water, eating foods contaminated by heavy metals, and living close to a location where these metals are improperly disposed of. These methods can have negative effects on people. Toxic effects of several HMs on human health are discussed in the following:

7.1. *Toxic effects of Cd*

One extremely hazardous heavy metal is cadmium. The US National Toxicology Program and the International Agency for Research on Cancer have designated cadmium as a human carcinogen (Congeevaram *et al.*, 2007). In addition to turning healthy epithelial cells into cancerous cells and making them carcinogenic, cadmium prevents the manufacture of DNA, RNA, and proteins. Long-term exposure to Cd in humans damages the kidneys and causes lung cancer. When Cd interacts with vital elements such as Fe, Mg, Ca, and Se, it alters the morphology and function of several organs and interferes with secondary metabolism. Pneumonia, overall weakness, fever, chest discomfort, and, in extreme situations, death, are all symptoms of Cd (Kiran *et al.*, 2021). Women are more impacted than males by the high concentration of Cd deposited in urine, blood, and the kidney cortex because of a rise in dietary Cd in the intestines of women itai-itai illness, which primarily affects women, is brought on by cadmium and results in many bone fractures from osteoporosis and osteomalacia (a condition in which calcium leaks from bones) (Cervantes *et al.*, 2001). When cadmium combines with hydrochloric acid in the stomach, it produces chlorides, which causes the gastrointestinal system to become acutely inflamed (Kao *et al.*, 2008). Because Cd primarily targets T cells (cytotoxic K (killer) cells), NK (natural killer cells), macrophages, and B cells, it results in immune system dysfunction (Mishra and Tripathi, 2009). The cardiovascular system becomes dysfunctional as a result of Cd's binding with metallothionein, a protein that controls zinc homeostasis and free radical scavenger. This leads to hypertension and an increase in reactive oxygen species (ROS) (Rodríguez *et al.*, 2007). When Cd is inhaled, respiratory distress syndrome results. By impairing the kidney's capacity to reabsorb vitamins, minerals, and nutrients, cadmium leads to renal abnormalities causing hypertension and diabetes. Chronic rhinitis, respiratory stress, and anosmia are caused by elevated Cd concentrations in the air. By disrupting testicular and prostate function, changing hormonal activity, and affecting male fertility, cadmium negatively impacts reproductive function (Choo *et al.*, 2006).

7.2. *Toxic effects of Ni*

Human health deteriorates when exposed to nickel. Ni impairs lung function, produces chronic bronchitis, and mostly causes lung cancer. Chronic exposure to carcinogenic Ni vapors, such as nickel oxide, resulted in sinus issues, asthma, and lung cancer. The kidney is the organ most impacted by nickel exposure. Frank haematuria, kidney damage, inflammation in bronchitis, lumen congestion, alveolar cell hyperplasia, carcinogenesis, chromosomal damage, mutagenesis, and inhibition of NK cell function due to Ni^{+2} acting as a tumour are all consequences of acute nickel poisoning (Pandi *et al.*, 2009). Ni increases the lipid (LPO) and causes reactive oxygen species (ROS) through iron accumulation. When iron levels rise, LPO lowers the amount of hepatic glutathione peroxidase, which leads to oxidative stress and peroxidative damage from hydroxyl radicals involving Haber–Weiss and Fenton chemistry (Florea and Büsselberg, 2006). Vertigo, sweating, palpitations, diarrhea, sleeplessness, and heaviness in the chest are all symptoms of Ni. Individual absorption of nickel ions via the skin in sensitized skin results in dermatitis, which manifests as skin irritation. In severe cases, it results in ulcers and pustules, and occasionally it produces skin eruptions and redness (Schaumlöffel, 2012). Ni is carcinogenic to humans because it builds up in the nucleus, mostly in the nucleolar fraction. Soluble Ni compounds are more powerful carcinogens than insoluble Ni compounds because soluble Ni particles leave the tissue more quickly, whereas insoluble Ni particles stay in the tissue longer (Poonkothai and Vijayavathi, 2012). Human lungs are impacted by Ni exposure because it causes oedema, bleeding, pulmonary fibrosis, degradation of the bronchial epithelium, and disorganized alveolar cells (Buxton *et al.*, 2019). Adult respiratory distress syndrome (ARDS) brought on by nickel can occasionally result in death.

7.3. *Toxic effects of As*

For humans, arsenic is a hazardous heavy metal. It is recognized that inorganic arsenic causes cancer. Exposure to low to moderate amounts of As results in neurological problems, diabetes, and liver and kidney problems. Women are thought to be more vulnerable to As-induced skin diseases than males, which can lead to dermatitis (Vahidnia *et al.*, 2007). Keratosis, melanosis, and pigmentation are skin lesions that are indicative of As exposure. Another organ that is susceptible to arsenic poisoning is the brain. Neurological problems are brought on by arsenic. More damage is done to neurones with long axons and sensory nerves than to those with short axons and motor nerves. Pain, numbness in the soles of the feet, and paraesthesia result from a loss in neurones' capacity to detoxify reactive oxygen species and glutathione synthesis (Kumar Mundey *et al.*, 2013). One of the main causes of neurotoxicity is oxidative stress (Risher, 2003). Premature birth, foetal loss in women, and uterine conception loss are all increased by arsenic. As promotes cellular hypertrophy, chronic bronchitis (inflammation of the respiratory system), and steatosis (fatty liver) (Kiran *et al.*, 2021). Since arsenic damages platelets, it causes cardiovascular disorders. Haematemesis (bloody vomiting), persistent stomach discomfort, weight loss, enlarged spleen, weakness, and irregular sleep patterns are all caused by arsenic. Ingestion of large levels of As can occasionally result in death.

7.4. *Toxic effects of Hg*

The most hazardous non-essential metal for humans is mercury. There are three types of mercury found in the environment: elemental, inorganic, and organic mercury. The primary way that elemental mercury is discharged into the atmosphere is as vapor (Kiran *et al.*, 2021). The central nervous system is the primary target of elemental mercury vapor, which causes cognitive, motor, and sensory disturbances with symptoms including memory loss, sleeplessness, tremors (affecting the hands and occasionally other areas of the body), and muscle twitches. Extended exposure to elemental mercury results in impaired focus, clouded vision, and shaky gait. Excessive mercury exposure damages the nervous system and can be fatal. Negative consequences of mercury that are transferred from the mother to the foetus include mental retardation, blindness, brain damage, and speech impairment. High levels of mercury exposure result in pulmonary oedema, pneumonia, and other signs of lung damage. Adults who are exposed to low levels of mercury experience sadness, tremors (shaking), skin rashes, memory loss, and children's hands and feet peeling off and becoming red (Rice *et al.*, 2014). About 50% of a dental filling is made of metallic mercury, and through amalgam surface abrasion, mercury is liberated from the amalgam as inorganic ions or mercury vapors. Mercury vapors in the mouth have the potential to enter the lungs through the respiratory system. The gastrointestinal system absorbs metal ions through oral fluids, while the circulation carries mercury vapors to various bodily organs (Wada *et al.*, 2009). Since it promotes oxidative stress and mitochondrial dysfunction, mercury raises lipid peroxidation and modifies calcium homeostasis. Since mercury acts as a catalyst for Fenton-type reactions, it raises the quantity of ROS. Fish consumption is the primary cause of organic mercury poisoning in humans. In Iraq and Japan, it causes Minimata illness. Dimethyl mercury, on the other hand, is more poisonous than methyl mercury; a few millilitres spilt on the skin can be fatal (Chen *et al.*, 2006). Since it blocks the formation of thyroid hormones and occupies iodine binding sites, mercury reduces the action of thyroid hormones and results in hypothyroidism, thyroid inflammation, and body temperature impairment (Mathew *et al.*, 2011). There is a tendency for the pancreas to accumulate mercury. Hg has three sulfur-binding sites for insulin, which are implicated in diabetes and lead to blood glucose dysregulation (Wadhwa *et al.*, 2012). Infertility in both males and females is also caused by mercury.

7.5. *Toxic effects of Pb*

Lead is regarded as a hazardous heavy metal. Pb has detrimental effects on the body even at low concentrations. Since Pb metal ions may substitute other monovalent cations such as Na^+ and bivalent cations such as

Fe^{+2}, Pb disrupts a variety of activities, including protein folding, inter- and intracellular signalling, apoptosis, enzyme control, and cell adhesion. In children aged 0–5, a lead deficiency has an impact on neurodevelopment. Since lead enters children through the placenta in the mother's uterus, exposure to lead during pregnancy raises major concerns. Since Pb raises ROS and lowers glutathione, an antioxidant that shields cells from free radicals such as hydrogen peroxide, it causes oxidative stress in live cells. Since ROS destroy proteins, cell membranes, and lipids structurally, they result in stressful conditions at the cellular level (Assi *et al.*, 2016). Lipid peroxidation is caused by free radicals stealing electrons from lipid molecules found in cell membranes (Vij, 2009). Lead can compete with calcium for binding and impact neural signals at picomolar concentrations on the binding sites of cerebellar phosphokinase C (Fang *et al.*, 2016). People who are exposed to high levels of lead develop hypertension and cardiovascular disease. It results in serious conditions such as ischaemic coronary heart disease, peripheral vascular disease, and cerebrovascular

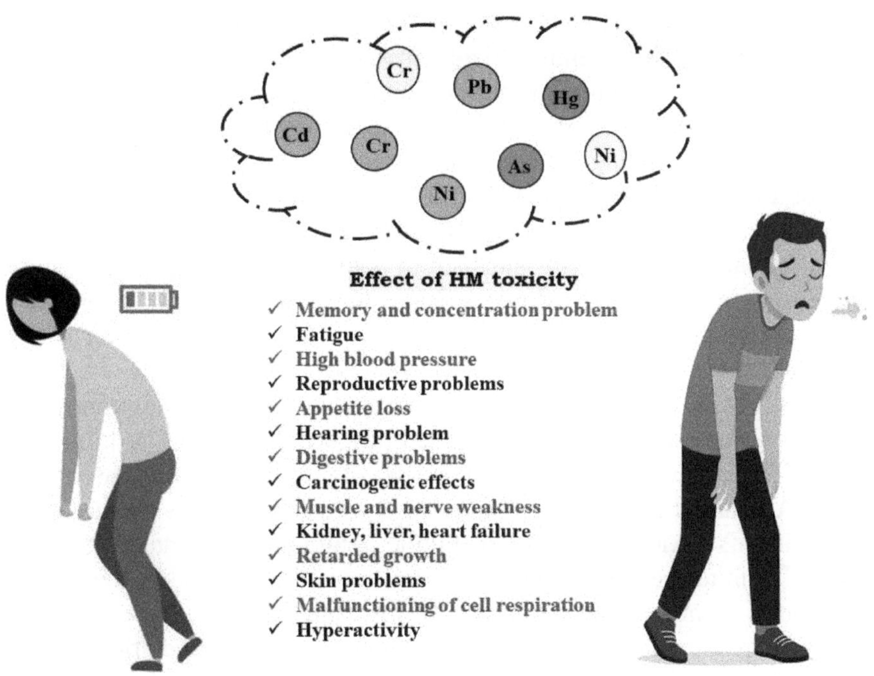

Fig. 4. Health problems associated with HM toxicity.

illness. Pb harms human kidneys as well. Pb produces two different forms of anaemia in humans: haemolytic anaemia, which is caused by high levels of lead exposure, and Frank anaemia, which is caused by prolonged elevated blood lead levels (Vij, 2009). Lead has an impact on both male and female reproductive systems. It causes spermatogenesis abnormalities, chromosomal damage, altered testosterone, decreased libido, and infertility in males; in females, it causes miscarriages, early membrane rupture, preterm delivery, and infertility.

Some health problems associated with HM toxicity are diagrammatically shown in Fig. 4.

8. Conclusion

The majority of heavy metals are hazardous and pollute the environment. Since heavy metals are water-soluble and non-biodegradable, they may readily penetrate plants and humans and have negative consequences. The ecosystem's natural and man-made activities are the primary sources of their discharge. This review presents the potential consequences of very toxic heavy metals, such as Cd, Ni, As, Hg, and Pb, on humans and plants in an understandable manner. Plants are impacted by heavy metals in a number of ways, including growth inhibition, a reduction in physiological and biological activity, and a variety of resistance to heavy metal detoxification. By influencing several human systems and functions, they also exhibit a wide range of impacts on the human body. As anticipated by the author, this chapter compiled all of the detrimental impacts of heavy metals on aquatic environment and human health in one location and may prove to be a valuable resource for future heavy metal study. In developing countries, one of the main sources of seafood contamination is the accumulation of heavy metals (HMs) in organisms. However, the main cause of seafood contamination in wealthy nations is the discharge of sludge or wastewater that has not been properly handled. The transmission of heavy metals is complex and involves several different procedures. A detailed assessment of a metal's precise toxicity is necessary to determine its toxicity in seafood. Public health concerns have been the subject of much research worldwide, but few of these studies have used efficient epidemiological methods. By lowering heavy metal concentrations in aquatic habitats and the food chain, current remediation procedures minimize health concerns. Rapid and accurate mapping of marine food contamination is necessary to stop metal pollutants from getting into the food chain and develop effective remediation methods. For moderately polluted water, biological treatment can be an economical and ecologically beneficial option. Nevertheless, no specific toxicity mechanism or remediation technique has been found. Therefore, creating practical, economical, and environmentally acceptable remediation techniques to eliminate these heavy metals from the environment will be one of our future problems.

Bibliography

Al-Ghanim, K.A., Mahboob, S., Seemab, S., Sultana, S., Sultana, T., Al-Misned, F., and Ahmed, Z. (2016). Monitoring of trace metals in tissues of Wallago attu (lanchi) from the Indus River as an indicator of environmental pollution. *Saudi Journal of Biological Sciences*, 23, 72–78. https://doi.org/10.1016/j.sjbs.2015.03.012.

Ali, H. and Khan, E. (2018). Bioaccumulation of non-essential hazardous heavy metals and metalloids in freshwater fish. Risk to human health. *Environmental Chemistry Letters*, 16, 903–917. https://doi.org/10.1007/s10311-018-0734-7.

Ali, H., Ali, W., Ullah, K., Akbar, F., Ahrar, S., Ullah, I., Ahmad, I., Ahmad, A., Ilahi, I., and Muhammad, A.S. (2017). Bioaccumulation of Cu and Zn in schizothorax plagiostomus and mastacembelus armatus from river swat, river panjkora and river Barandu in Malakand division, Pakistan. *Pakistan Journal of Zoology*, 49, 1555–1561. https://doi.org/10.17582/journal.pjz/2017.49.5.1555.1561.

Ali, H., Khan, E., and Ilahi, I. (2019). Environmental chemistry and ecotoxicology of hazardous heavy metals: Environmental persistence, toxicity, and bioaccumulation. *Journal of Chemistry*, 2019, 6730305. https://doi.org/10.1155/2019/6730305.

Alizada, N., Malik, S., and Bin Muzaffar, S. (2020). Bioaccumulation of heavy metals in tissues of Indian anchovy (Stolephorus indicus) from the UAE coast, Arabian Gulf. *Marine Pollution Bulletin*, 154, 111033. https://doi.org/10.1016/j.marpolbul.2020.111033.

Atti, S.K., Silver, E.M., Chokshi, Y., Casteel, S., Kiernan, E., dela Cruz, R., Kazzi, Z., and Geller, R.J. (2020). All that glitters is not gold: Mercury poisoning in a family mimicking an infectious illness. *Current Problems in Pediatric and Adolescent Health Care*, 50, 100758. https://doi.org/10.1016/j.cppeds.2020.100758.

Assi, M.A., Hezmee, M.N.M., Haron, A.W., Sabri, M.Y.M., and Rajion, M.A. (2016). The detrimental effects of lead on human and animal health.

Veterinary World, 9, 660–671. https://doi.org/10.14202/vetworld.2016.660-671.

Barbee, J.Y. and Prince, T.S. (1999). Acute respiratory distress syndrome in a welder exposed to metal fumes. *Southern Medical Journal*, 92, 510–512. https://doi.org/10.1097/00007611-199905000-00012.

Bernhoft, R.A. (2012). Mercury toxicity and treatment: A review of the literature. *Journal of Environmental and Public Health*, 2012, 460508. https://doi.org/10.1155/2012/460508.

Blaser, P., Zimmermann, S., Luster, J., and Shotyk, W. (2000). Critical examination of trace element enrichments and depletions in soils: As, Cr, Cu, Ni, Pb, and Zn in Swiss forest soils. *Science of the Total Environment*, 249, 257–280. https://doi.org/10.1016/S0048-9697(99)00522-7.

Bonsignore, M., Salvagio Manta, D., Mirto, S., Quinci, E.M., Ape, F., Montalto, V., Gristina, M., Traina, A., and Sprovieri, M. (2018). Bioaccumulation of heavy metals in fish, crustaceans, molluscs and echinoderms from the Tuscany coast. *Ecotoxicology and Environmental Safety*, 162, 554–562. https://doi.org/10.1016/j.ecoenv.2018.07.044.

Brodin, M., Vallejos, M., Opedal, M.T., Area, M.C., and Chinga-Carrasco, G. (2017). Lignocellulosics as sustainable resources for production of bioplastics – A review. *Journal of Cleaner Production*, 162, 646–664. https://doi.org/10.1016/j.jclepro.2017.05.209.

Buxton, S., Garman, E., Heim, K.E., Lyons-Darden, T., Schlekat, C.E., Taylor, M.D., and Oller, A.R. (2019). Concise review of nickel human health toxicology and ecotoxicology. *Inorganics*, 7, 89. https://doi.org/10.3390/inorganics7070089.

Casado-Martinez, M.C., Smith, B.D., Luoma, S.N., and Rainbow, P.S. (2010). Bioaccumulation of arsenic from water and sediment by a deposit-feeding polychaete (Arenicola marina): A biodynamic modelling approach. *Aquatic Toxicology*, 98, 34–43. https://doi.org/10.1016/j.aquatox.2010.01.015.

Castro-González, M.I. and Méndez-Armenta, M. (2008). Heavy metals: Implications associated to fish consumption. *Environmental Toxicology and Pharmacology*, 26, 263–271. https://doi.org/10.1016/j.etap.2008.06.001.

Cervantes, C., Campos-García, J., Devars, S., Gutiérrez-Corona, F., Loza-Tavera, H., Torres-Guzmán, J.C., and Moreno-Sánchez, R. (2001). Interactions of chromium with microorganisms and plants. *FEMS Microbiology Reviews*, 25, 335–347. https://doi.org/10.1016/S0168-6445(01)00057-2.

Chakraborty, P., Prithiviraj, B., Selvaraj, S., and Kumar, B. (2016). Polychlorinated biphenyls in settled dust from informal electronic waste recycling workshops and nearby highways in urban centers and suburban industrial roadsides of Chennai city, India: Levels, congener profiles and exposure assessment. *Science of the Total Environment*, 573, 1413–1421. https://doi.org/10.1016/j.scitotenv.2016.07.129.

Chen, Y.W., Huang, C.F., Tsai, K.S., Yang, R.S., Yen, C.C., Yang, C.Y., Lin-Shiau, S.Y., and Liu, S.H. (2006). Methylmercury induces pancreatic β-cell apoptosis and dysfunction. *Chemical Research in Toxicology*, 19, 1080–1085. https://doi.org/10.1021/tx0600705.

Choo, T.P., Lee, C.K., Low, K.S., and Hishamuddin, O. (2006). Accumulation of chromium (VI) from aqueous solutions using water lilies (Nymphaea spontanea). *Chemosphere*, 62, 961–967. https://doi.org/10.1016/j.chemosphere.2005.05.052.

Congeevaram, S., Dhanarani, S., Park, J., Dexilin, M., and Thamaraiselvi, K. (2007). Biosorption of chromium and nickel by heavy metal resistant fungal and bacterial isolates. *Journal of Hazardous Materials*, 146, 270–277. https://doi.org/10.1016/j.jhazmat.2006.12.017.

Corrias, F., Atzei, A., Addis, P., Secci, M., Russo, M., and Angioni, A. (2020). Integrated environmental evaluation of heavy metals and metalloids bioaccumulation in invertebrates and seaweeds from different marine coastal areas of sardinia, mediterranean sea. *Environmental Pollution*, 266, 115048. https://doi.org/10.1016/j.envpol.2020.115048.

Dane, H. and Şişman, T. (2020). A morpho-histopathological study in the digestive tract of three fish species influenced with heavy metal pollution. *Chemosphere*, 242, 125212. https://doi.org/10.1016/j.chemosphere.2019.125212.

Dieter, H.H., Bayer, T.A., and Multhaup, G. (2005). Environmental copper and manganese in the pathophysiology of neurologic diseases (Alzheimer's disease and manganism). *Acta Hydrochimica et Hydrobiologica*, 33, 72–78. https://doi.org/10.1002/aheh.200400556.

Duffus, J.H. (2002). "Heavy metals" – a meaningless term? *Pure and Applied Chemistry*, 74, 793–807. https://doi.org/10.1351/pac200274050793.

El-Moselhy, K.M., Othman, A.I., Abd El-Azem, H., and El-Metwally, M.E.A. (2014). Bioaccumulation of heavy metals in some tissues of fish in the Red Sea, Egypt. *Egyptian Journal of Basic and Applied Sciences*, 1, 97–105. https://doi.org/10.1016/j.ejbas.2014.06.001.

Fang, J., Lang, M., Ye, X., Zhang, W., and Zhu, K. (2016). Non-isothermal crystallization behavior of polypropylene/zinc oxide composites. *Science and Engineering of Composite Materials*, 23, 505–510. https://doi.org/10.1515/secm-2014-0078.

Ferrey, M.L., Hamilton, M.C., Backe, W.J., and Anderson, K.E. (2018). Pharmaceuticals and other anthropogenic chemicals in atmospheric particulates and precipitation. *Science of the Total Environment*, 612, 1488–1497. https://doi.org/10.1016/j.scitotenv.2017.06.201.

Florea, A.-M. and Büsselberg, D. (2006). Occurrence, use and potential toxic effects of metals and metal compounds. *Biometals*, 19, 419–427. https://doi.org/10.1007/s10534-005-4451-x.

Gouws, G. and Stewart, B.A. (2001). Potamonautid river crabs (Decapoda, Brachyura, Potamonautidae) of KwaZulu-Natal, South Africa. *Water SA*, 27, 85–98. https://doi.org/10.4314/wsa.v27i1.5015.

Govind, P. and Madhuri, S. (2014). Heavy metals causing toxicity in humans, animals and environment. *Research Journal of Animal, Veterinary and Fishery Sciences*, 2, 17–23.

Guzzi, G., Ronchi, A., and Pigatto, P. (2021). Toxic effects of mercury in humans and mammals. *Chemosphere*, 263, 127990. https://doi.org/10.1016/j.chemosphere.2020.127990.

Harrison, R. and Alloway, B. (2007). Soil pollution and land contamination. In: *Pollution: Causes, Effects and Control*. Royal Society of Chemistry, pp. 352–377. https://doi.org/10.1039/9781847551719-00352.

Jardine, T.D., Galloway, A.W.E., and Kainz, M.J. (2020). Unlocking the power of fatty acids as dietary tracers and metabolic signals in fishes and aquatic invertebrates. *Philosophical Transactions of the Royal Society B: Biological Sciences*, 375, 20190639. https://doi.org/10.1098/rstb.2019.0639.

Jezierska, B., Ługowska, K., and Witeska, M. (2009). The effects of heavy metals on embryonic development of fish (a review). *Fish Physiology and Biochemistry*, 35, 625–640. https://doi.org/10.1007/s10695-008-9284-4.

Juma, H., Battah, A., Salim, M., and Tiwari, P. (2002). Arsenic and cadmium levels in imported fresh and frozen fish in jordan. *Bulletin of Environmental Contamination and Toxicology*, 68, 132–137. https://doi.org/10.1007/s00128-001-0229-1.

Kao, W.-C., Huang, C.-C., and Chang, J.-S. (2008). Biosorption of nickel, chromium and zinc by MerP-expressing recombinant Escherichia coli. *Journal of Hazardous Materials*, 158, 100–106. https://doi.org/10.1016/j.jhazmat.2008.01.032.

Khan, M.N., Mobin, M., Abbas, Z.K., and Alamri, S.A. (2017). Fertilizers and their contaminants in soils, surface and groundwater. Elsevier Inc. https://doi.org/10.1016/B978-0-12-809665-9.09888-8.

Khatun, J., Intekhab, A., and Dhak, D. (2022). Effect of uncontrolled fertilization and heavy metal toxicity associated with arsenic(As), lead(Pb) and cadmium (Cd), and possible remediation. *Toxicology*, 477, 153274. https://doi.org/10.1016/j.tox.2022.153274.

Kiran, Bharti, R., and Sharma, R. (2021). Effect of heavy metals: An overview. *Materials Today: Proceedings*, 51, 880–885. https://doi.org/10.1016/j.matpr.2021.06.278.

Knight, C., Kaiser, J., Lalor, G.C., Robotham, H., and Witter, J.V. (1997). Heavy metals in surface water and stream sediments in Jamaica. *Environmental Geochemistry and Health*, 19, 63–66. https://doi.org/10.1023/A:1018442219943.

Kodama, K., Lee, J.-H., Oyama, M., Shiraishi, H., and Horiguchi, T. (2012). Disturbance of benthic macrofauna in relation to hypoxia and organic enrichment in a eutrophic coastal bay. *Marine Environmental Research*, 76, 80–89. https://doi.org/10.1016/j.marenvres.2011.08.007.

Krieger, J.R., Beaudreau, A.H., Heintz, R.A., and Callahan, M.W. (2020). Growth of young-of-year sablefish (Anoplopoma fimbria) in response to temperature and prey quality: Insights from a life stage specific bioenergetics model. *Journal of Experimental Marine Biology and Ecology*, 526, 151340. https://doi.org/10.1016/j.jembe.2020.151340.

Kucukosmanoglu, A.G. and Filazi, A. (2020). Investigation of the metal pollution sources in Lake Mogan, Ankara, Turkey. *Biological Trace Element Research*, 198, 269–282. https://doi.org/10.1007/s12011-020-02031-z.

Kumar Mundey, M., Roy, M., Roy, S., Kumar Awasthi, M., and Sharma, R. (2013). Antioxidant potential of Ocimum sanctum in arsenic induced nervous tissue damage. *Brazilian Journal of Veterinary Pathology*, 6, 95–101.

Kumar, M., Gupta, N., Ratn, A., Awasthi, Y., Prasad, R., Trivedi, A., and Trivedi, S.P. (2020). Biomonitoring of heavy metals in River Ganga water, sediments, plant, and fishes of different trophic levels. *Biological Trace Element Research*, 193, 536–547. https://doi.org/10.1007/s12011-019-01736-0.

Landolph, J.R. (1994). Molecular mechanisms of transformation of C3H/10T1/2 C1 8 mouse embryo cells and diploid human fibroblasts by carcinogenic metal compounds. *Environmental Health Perspectives*, 102, 119–125. https://doi.org/10.1289/ehp.94102s3119.

Liu, M., Xu, Y., Nawab, J., Rahman, Z., Khan, S., Idress, M., Ud din, Z., Ali, A., Ahmad, R., Khan, S.A., Khan, A., Khan, M.Q., Tang, Y.-T., and Li, G. (2020). Contamination features, geo-accumulation, enrichments and human health risks of toxic heavy metal(loids) from fish consumption collected along Swat river, Pakistan. *Environmental Technology & Innovation*, 17, 100554. https://doi.org/10.1016/j.eti.2019.100554.

Maheswari Nair K. V. Jayalakshmy, K.K.B. and Joseph, T. (2006). Bioaccumulation of toxic metals by fish in a semi-enclosed tropical ecosystem. *Environmental Forensics*, 7, 197–206. https://doi.org/10.1080/15275920600840438.

Mansouri, B., Ebrahimpour, M., and Babaei, H. (2012). Bioaccumulation and elimination of nickel in the organs of black fish (Capoeta fusca). *Toxicology and Industrial Health*, 28, 361–368. https://doi.org/10.1177/0748233711412425.

Mathew, B.B., Tiwari, A., and Jatawa, S.K. (2011). Free radicals and antioxidants: A review. *Journal of Pharmacy Research*, 4, 4340–4343.

Maurya, P.K., Malik, D.S., Yadav, K.K., Kumar, A., Kumar, S., and Kamyab, H. (2019). Bioaccumulation and potential sources of heavy metal contamination in fish species in River Ganga basin: Possible human health risks evaluation. *Toxicology Reports*, 6, 472–481. https://doi.org/10.1016/j.toxrep.2019.05.012.

Mehmood, M.A., Qadri, H., Bhat, R.A., Rashid, A., Ganie, S.A., Dar, G.H., and Shafiq-ur-Rehman. (2019). Heavy metal contamination in two commercial fish species of a trans-Himalayan freshwater ecosystem. *Environmental Monitoring and Assessment*, 191, 104. https://doi.org/10.1007/s10661-019-7245-2.

Mishra, V.K. and Tripathi, B.D. (2009). Accumulation of chromium and zinc from aqueous solutions using water hyacinth (Eichhornia crassipes). *Journal of Hazardous Materials*, 164, 1059–1063. https://doi.org/10.1016/j.jhazmat.2008.09.020.

Moiseenko, T.I. and Gashkina, N.A. (2020). Distribution and bioaccumulation of heavy metals (Hg, Cd and Pb) in fish: Influence of the aquatic environment and climate. *Environmental Research Letters*, 15, 115013. https://doi.org/10.1088/1748-9326/abbf7c.

Mortuza, M.G. and Al-Misned, F.A. (2015). Heavy metal concentration in two freshwater fishes from Wadi Hanifah (Riyadh, Saudi Arabia) and evaluation of possible health hazard to consumers. *Pakistan Journal of Zoology*, 47, 839–845.

Muhammad, S., Shah, M.T., and Khan, S. (2011). Health risk assessment of heavy metals and their source apportionment in drinking water of Kohistan region, northern Pakistan. *Microchemical Journal*, 98, 334–343. https://doi.org/10.1016/j.microc.2011.03.003.

Olaifa, F., Olaifa, A., and Onwude, T. (2010). Lethal and sub-lethal effects of copper to the african catfish (*Clarias gariepinus*) juveniles. *African Journal of Biomedical Research*, 7, 65–70. https://doi.org/10.4314/ajbr.v7i2.54071.

Ouyang, Y., Higman, J., Thompson, J., O'Toole, T., and Campbell, D. (2002). Characterization and spatial distribution of heavy metals in sediment from Cedar and Ortega rivers subbasin. *Journal of Contaminant Hydrology*, 54, 19–35. https://doi.org/10.1016/S0169-7722(01)00162-0.

Oze, G.O., Oze, R.N., Ananuso, C.I., Ogukwe, C., and Nwanjo, H. (2009). Metal pollution of fish of Qua-Iboe River Estuary. *Journal of Applied Sciences and Environmental Management*, 24, 1–5.

Pandi, M., Shashirekha, V., and Swamy, M. (2009). Bioabsorption of chromium from retan chrome liquor by cyanobacteria. *Microbiological Research*, 164, 420–428. https://doi.org/10.1016/j.micres.2007.02.009.

Poonkothai, M. and Vijayavathi, B.S. (2012). Nickel as an essential element and a toxicant. *International Journal of Environmental Sciences*, 1, 285–288.

Pragnya, M., Dinesh Kumar, S., Solomon Raju, A.J., and Murthy, L.N. (2020). Bioaccumulation of heavy metals in different organs of Labeo rohita, Pangasius hypophthalmus, and Katsuwonus pelamis from Visakhapatnam, India. *Marine Pollution Bulletin*, 157, 111326. https://doi.org/10.1016/j.marpolbul.2020.111326.

Rai, P.K., Lee, S.S., Zhang, M., Tsang, Y.F., and Kim, K.-H. (2019). Heavy metals in food crops: Health risks, fate, mechanisms, and management. *Environment International*, 125, 365–385. https://doi.org/10.1016/j.envint.2019.01.067.

Rice, K.M., Walker, E.M., Wu, M., Gillette, C., and Blough, E.R. (2014). Environmental mercury and its toxic effects. *Journal of Preventive Medicine and Public Health*, 47, 74–83. https://doi.org/10.3961/jpmph.2014.47.2.74.

Risher, J.F. (2003). *Elemental Mercury and Inorganic Mercury Compounds: human Health Aspects*. World Health Organization, Geneva, pp. iv + 61.

Rodríguez, M.C., Barsanti, L., Passarelli, V., Evangelista, V., Conforti, V., and Gualtieri, P. (2007). Effects of chromium on photosynthetic and photoreceptive apparatus of the alga Chlamydomonas reinhardtii. *Environmental Research*, 105, 234–239. https://doi.org/10.1016/j.envres.2007.01.011.

Saher, N.U. and Siddiqui, A.S. (2019). Occurrence of heavy metals in sediment and their bioaccumulation in sentinel crab (Macrophthalmus depressus) from highly impacted coastal zone. *Chemosphere*, 221, 89–98. https://doi.org/10.1016/j.chemosphere.2019.01.008.

Sankhla, M.S. (2019). Contaminant of heavy metals in groundwater & its toxic effects on human health & environment. *International Journal of Environmental Sciences & Natural Resources*, 18, 555996. https://doi.org/10.19080/ijesnr.2019.18.555996.

Savci, S. (2012). Investigation of effect of chemical fertilizers on environment. *APCBEE Procedia*, 1, 287–292. https://doi.org/10.1016/j.apcbee.2012.03.047.

Schaumlöffel, D. (2012). Nickel species: Analysis and toxic effects. *Journal of Trace Elements in Medicine and Biology*, 26, 1–6. https://doi.org/10.1016/j.jtemb.2012.01.002.

Senthil Murugan, S., Karuppasamy, R., Poongodi, K., and Puvaneswari, S. (2008). Bioaccumulation pattern of zinc in freshwater fish Channa punctatus (Bloch.) after chronic exposure. *Turkish Journal of Fisheries and Aquatic Sciences*, 59, 55–59.

Sharma, N. and Singhvi, R. (2017). Effects of chemical fertilizers and pesticides on human health and environment: A review. *International Journal of Agriculture, Environment and Biotechnology*, 10, 675. https://doi.org/10.5958/2230-732x.2017.00083.3.

Sonone, S.S., Jadhav, S., Sankhla, M.S., and Kumar, R. (2021). Water contamination by heavy metals and their toxic effect on aquaculture and human health through food chain. *Letters in Applied NanoBioScience*, 10, 2148–2166. https://doi.org/10.33263/LIANBS102.21482166.

Squadrone, S., Brizio, P., Stella, C., Mantia, M., Favaro, L., Biancani, B., Gridelli, S., Da Rugna, C., and Abete, M.C. (2020). Differential bioaccumulation of trace elements and rare earth elements in the muscle, kidneys, and liver of the invasive Indo-Pacific lionfish (Pterois spp.) from Cuba. *Biological Trace Element Research*, 196, 262–271. https://doi.org/10.1007/s12011-019-01918-w.

Srivastava, V., de Araujo, A.S.F., Vaish, B., Bartelt-Hunt, S., Singh, P., and Singh, R.P. (2016). Biological response of using municipal solid waste compost in agriculture as fertilizer supplement. *Reviews in Environmental Science and Bio/Technology*, 15, 677–696. https://doi.org/10.1007/s11157-016-9407-9.

Srivastava, V., Gupta, S.K., Singh, P., Sharma, B., and Singh, R.P. (2018). Biochemical, physiological, and yield responses of lady's finger (Abelmoschus esculentus L.) grown on varying ratios of municipal solid waste vermicompost. *International Journal of Recycling of Organic Waste in Agriculture*, 7, 241–250. https://doi.org/10.1007/s40093-018-0210-1.

Strachan, S. (2010). Trace elements. *Current Anaesthesia & Critical Care*, 21, 44–48. https://doi.org/10.1016/j.cacc.2009.08.004.

Suter, G. and Suter, G. (2009). Environments: Science and lateral. *Environmental Toxicology and Chemistry*, 5, 492–493.

Topal, T. and Onac, C. (2020). Determination of heavy metals and pesticides in different types of fish samples collected from four different locations of Aegean and Marmara Sea. *Journal of Food Quality*, 2020, 8101532. https://doi.org/10.1155/2020/8101532.

Vahidnia, A., van der Voet, G.B., and de Wolff, F.A. (2007). Arsenic neurotoxicity – A review. *Human & Experimental Toxicology*, 26, 823–832. https://doi.org/10.1177/0960327107084539.

Vij, A.G. (2009). Hemopoietic, hemostatic and mutagenic effects of lead and possible prevention by zinc and vitamin C. *Al Ameen Journal of Medical Sciences*, 2, 27–36.

Vosylienė, M. and Jankaitė, A. (2006). Effect of heavy metal model mixture on rainbow trout biological parameters. *Ekologija*, 52, 12–17.

Wada, H., Cristol, D.A., McNabb, F.M.A., and Hopkins, W.A. (2009). Suppressed adrenocortical responses and thyroid hormone levels in birds near a mercury-contaminated river. *Environmental Science & Technology*, 43, 6031–6038. https://doi.org/10.1021/es803707f.

Wadhwa, N., Mathew, B.B., Jatawa, S.K., and Tiwari, A. (2012). Lipid peroxidation: mechanism, models and significance. *International Journal of Current Science*, 3, 11–17.

Wasserman, W.G.A., Xinhua, L., Faruque, P., Habibul, A., Diane, L., Pam, F.-L., Jennie, K., Alexander, van G., Vesna, S., J., L.N., Zhongqi, C., Yan, Z., and H., G.J. (2006). Water manganese exposure and children's intellectual function in Araihazar, Bangladesh. *Environmental Health Perspectives*, 114, 124–129. https://doi.org/10.1289/ehp.8030.

Yogeshwaran, A., Gayathiri, K., Muralisankar, T., Gayathri, V., Monica, J.I., Rajaram, R., Marimuthu, K., and Bhavan, P.S. (2020). Bioaccumulation of heavy metals, antioxidants, and metabolic enzymes in the crab Scylla serrata from different regions of Tuticorin, Southeast Coast of India. *Marine Pollution Bulletin*, 158, 111443. https://doi.org/10.1016/j.marpolbul.2020.111443.

Yuan, Y., Sun, T., Wang, H., Liu, Y., Pan, Y., Xie, Y., Huang, H., and Fan, Z. (2020). Bioaccumulation and health risk assessment of heavy metals to bivalve species in Daya Bay (South China Sea): Consumption advisory. *Marine Pollution Bulletin*, 150, 110717. https://doi.org/10.1016/j.marpolbul.2019.110717.

Zaynab, M., Al-Yahyai, R., Ameen, A., Sharif, Y., Ali, L., Fatima, M., Khan, K.A., and Li, S. (2022). Health and environmental effects of heavy metals. *Journal of King Saud University - Science*, 34, 101653. https://doi.org/10.1016/j.jksus.2021.101653.

Zhang, W., Xue, J., Ge, M., Yu, M., Liu, L., and Zhang, Z. (2013). Resveratrol attenuates hepatotoxicity of rats exposed to arsenic trioxide. *Food and Chemical Toxicology*, 51, 87–92. https://doi.org/10.1016/j.fct.2012.09.023.

Ziyaadini, M., Yousefiyanpour, Z., Ghasemzadeh, J., and Zahedi, M.M. (2017). Biota-sediment accumulation factor and concentration of heavy metals (Hg, Cd, As, Ni, Pb and Cu) in sediments and tissues of Chiton lamyi (Mollusca: Polyplacophora: Chitonidae) in Chabahar Bay, Iran. *Iranian Journal of Fisheries Sciences*, 16. Available at: http://jifro.ir/article-1-2952-en.html.

https://doi.org/10.1142/9789819822249_0006

Chapter 6

Risk Assessment of Hazardous Emissions from Landfill Fire on the Ecosystem and Human Health

Chibuisi Gideon Alimba*,‡ and Samuel Oluwasegun Adesida†

**IU Internationale Hochschule, Buchdorf, Germany*

†Department of Environmental Protection in Agriculture, Faculty of Agriculture, University of Belgrade, Belgrade, Serbia.

‡chivoptera@yahoo.com

Abstract

Globally, anthropogenic (including industrial, agricultural, medical, municipal, and educational) activities generate myriads of solid wastes, which are capable of emitting thousands of hazardous chemicals, when decomposed or combusted, into the environment and adversely impact human health. In many countries, inappropriate disposal of these wastes into unsanitary landfills and setting them on fire to reduce waste volume is common, and this act increases hazardous contaminants in the environment (soil and air pollution) and can elicit deleterious health consequences on human and wildlife. In most cases, landfill fires are uncontrolled and generate a huge number of pollutants, including particulate matters, carbon monoxide, sulphur oxides, oxides of nitrogen, volatile organic compounds, benzene, polycyclic aromatic hydrocarbons, polychlorinated dibenzo-*p*-dioxins and dibenzofurans, and polychlorinated biphenyls, as well as some trace elements, such as arsenic, cadmium, nickel, zinc, mercury, chromium, and vanadium, are also generated. The pollutants generated from landfill fires have been reported to increase air pollution and soil degradation. Some of the gases are sources of greenhouse gas emissions and significantly exacerbate global climate change. While many others cause threat to human and wildlife health, including initiating and increasing respiratory, immunological, and dermal problem that potentially leads to more serious health outcomes, such as cancer. This chapter discusses the cause of landfill fire, environmental contamination of its emitted pollutants, human exposure pathways, and mechanisms of health consequences. It also suggests possible mitigation and the role of government and individuals in solid waste management to avert adverse outcomes of solid waste management.

Keywords: Landfill fire pollutants, environmental contamination, human exposure pathway, risk of health impact.

1. Introduction

Solid wastes are a product of human activities, including industrial, agricultural, municipal, and medical activities. With growing human populations and increasing advancements in science and technology, solid wastes are unavoidably generated in an attempt to satisfy ever-growing human needs, and these wastes are increasing at an unprecedented rate worldwide (Alimba, 2013). Therefore, the management strategies for the increasingly generated solid wastes are becoming almost unachievable and uncontrollable. Hence, solid waste accumulation in the environment is heightened (El-Fadel *et al.*, 1997; Alimba, 2013; Alimba *et al.*, 2021). The constituents of solid wastes are mainly textile, metals, glass, plastic, food wastes, rubber, wood, tires, paper and paperboard products, yard trimmings and other vegetations, electrical and electronics, and other materials in the solid states which are no longer useful to the person(s) or organizations generating them (Alimba, 2013; Alimba *et al.*, 2021). Solid wastes are mostly classified according to their source of generation and level of hazardousness. For instance, classification according to sources of generation may include the following:

(i) **Domestic wastes:** solid wastes generated during domestic activities, including from home kitchen, cooking centres, and human raw faeces.

(ii) **Agricultural wastes:** solid wastes generated from animal husbandries; animal beddings and feeds, excreta, carcasses, slaughtering and farming activities, including agrochemical and fertilizer applications, and pre- and post-harvest activities.

(iii) **Industrial wastes:** solid wastes generated during manufacturing activities, distribution of goods, and rendering of services. These waste types include metal scraps, chips, and grits from machine shops, sawdust, waste papers, pieces of glass, tyres, abandoned vehicles, metals, paper and pulp, iron and steel, petroleum sludge, textiles, etc.

(iv) **Radioactive wastes:** solid waste type generated from nuclear weapons during warfare or accidental discharge or illegal dumping. Radionuclide wastes, including uranium.

(v) **Medical (hospital, health care, or clinical) wastes:** solid wastes generated from human and/or animal health care activities, including research, diagnosis, and therapeutic and rehabilitative services or prevention. Examples of medical solid wastes include culture media and stock of infectious agents, pathological wastes, wastes from surgery or autopsy, sharp objects (hypodermal needles, syringes with needles, and scalpel blades), clotted blood, carcasses and tissue-related swabs, broken glassware, post mortems, pharmaceutical products, bedpans, incontinence pads, sanitary towels, and animal drugs.

(vi) **Waste Electrical and Electronic Equipment (WEEE or E-wastes):** solid wastes generated from all secondary computers, entertainment device electronics, mobile phones, television sets and refrigerators, monitors, batteries, and stereos, polyvinyl chloride (PVC plastics), etc.

Many of the solid wastes listed above are described as hazardous wastes since they pose substantial or potential threats to public health and the environment. These wastes are capable of exhibiting ignitability (flammability), reactivity, corrosivity, and toxicity (carcinogenicity, mutagenicity, teratogenicity, and systemic toxicity), which are dangerous to the environment and human health and other living things. Hazardous solid wastes are usually treated before disposal into designated sanitary facilities for hazardous waste facilities. These are readily available in many developed countries. On the contrary, in many developing countries, solid wastes are not usually treated due to the high cost of treatment, and they are co-disposed with municipal solid wastes in unsanitary landfills (Alimba 2013; Alimba *et al.*, 2021). This usually results in unfavorable discharge of hazardous chemicals (present in leachates and gases) into the environment (Alimba *et al.*, 2006; Alimba *et al.*, 2012, 2015, 2016, 2021, 2022a; Morales *et al.*, 2018).

2. Methods of Managing Solid Wastes

Generated huge quantity of solid waste in the landfill (environment) may constitute eyesores, serving as breeding sites and hiding places for insects (flies, mosquitoes, etc.), rodents (rats, mice, etc.), reptiles (snakes, lizards, etc.), and other predators. Occasionally, mostly in developing countries, stray animals, including dogs, cattle, sheep, goats, and human scavengers, are attracted to garbage sites for food and picking of possible useful materials. Activities of these animals and humans in and around the landfill sites are capable of increasing cases of diseases (including dysentery, malaria, dengue, yellow fever, and micro-organism-caused diseases). Snakes and other poisonous animals in the landfills can cause health deterioration and death to humans and domestic animals around the landfill facilities. Solid wastes in landfills can be washed by runoff water during rainfall to contaminate underground and surface water resources. Also, solid wastes can block waterways and channels, leading to flooding. Furthermore, Heaps of solid waste can cause fire outbreaks and smoke laden with particulate matter and volatile organic compounds, which can constitute a health hazard. Exposure to components of landfill ashes and smokes can increase breathing difficulties, neurological disorders, and other related health and environmental challenges.

Appropriate and adequate solid waste management will prevent the piling up of solid wastes in the environment. Also, it will prevent possible harmful effects on humans and other living creatures and reduce environmental contamination. Managing solid waste is a complex process that is beyond just disposal of the wastes in a specific place or location, but it includes the collection, processing, and transporting systems of the generated wastes, as well as minimization of wastes at the point of generation. Some of the common ways of managing solid wastes around the world include the following:

(i) reduction at source,
(ii) re-use and recycling,

(iii) composting,
(iv) incineration,
(v) landfilling,
(vi) pyrolysis/gasification,
(vii) anaerobic digestion.

Standard practices of the above-listed solid waste management methods are expensive to institute and may require highly skilled technical knowledge to operate. This makes it difficult to be readily practiced in many countries (El-Fadel *et al.*, 1997).

3. Solid Waste Landfilling: The Most Common Methods of Solid Waste Management

Landfilling is the dumping of solid wastes on excavated land. Excavated land describes a wide spectrum of sites ranging from managed, engineered, regulated to illegal sites, uncontrolled and indiscriminate dumping of wastes in the environment (El-Fadel *et al.*, 1997; Alimba, 2013, 2024). Landfilling is the most common method of managing solid wastes in most countries around the world (Yaashikaa *et al.*, 2022; Alimba, 2024). In many developed countries, a typical municipal landfill is an engineered site with the base lined with impermeable clay, plastic, rubber, or a composite layer covered by earth. At the end of each day's activity on the sites, wastes are covered with inert materials such as clay soil, and when the cell is eventually filled up, it is covered over with a layer of inert materials. During operation, a fence is built around the site to prevent wind from blowing materials off the site. A drainage system is built to collect wastewater runoff and leachates (see Fig. 1 for an example of an engineered landfill). Also, an energy recovery system is constructed to collect gas(es), which can either be used to generate electricity or is flared. Furthermore, landfill operations are regulated to protect human health. These regulations include banning the disposal of liquids, infectious clinical wastes, tires, and the co-disposal of hazardous wastes together with municipal wastes into the same landfills.

Solid waste disposal into landfills has some benefits which encourage individuals, organizations, and the government to readily dispose of waste into landfills. These benefits include the following:

(i) The cost of disposal is relatively cheap when using non-engineered landfills.

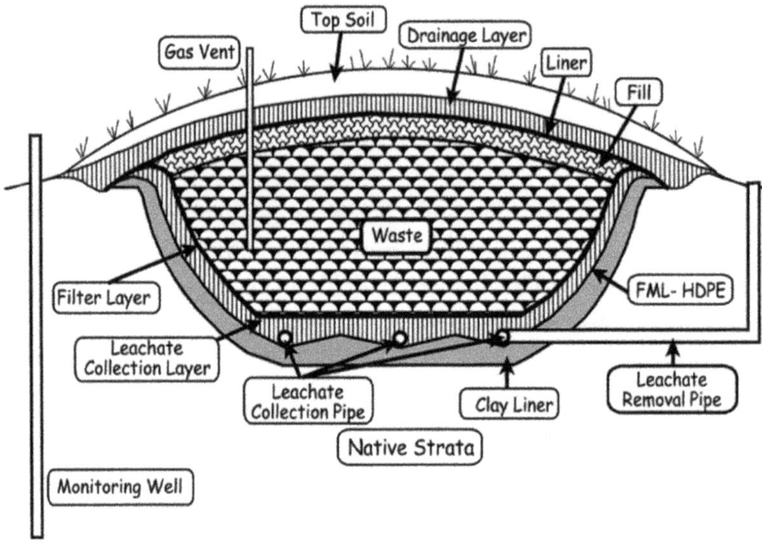

Fig. 1. A typical sanitary landfill with liner systems at the base and sides to provide barriers that minimize migration of contaminants from the containment site to groundwater. It has an engineered capping system, which prevents or controls infiltration of precipitation, minimizing movement of fluids through the wastes. It contains leachate collection systems that remove all leachate generated into leachate wells, where they are treated or attenuated. *Source*: Lee and Jones (1991).

(ii) Disposing or dumping of solid wastes into disused quarries and abandoned industrial sites is used to backfill excavated land before reclamation.
(iii) When such a landfill is filled up and the land is not meant for reclamation, the wastes are burned to reduce waste volume; hence, dumping of wastes continues.

Unfortunately, the above benefits generate devastating disadvantages which include the following:

(i) The production of leachates and runoff waste water greatly contaminates the surrounding soil, underground, and surface water and air of the operating landfill facilities. Furthermore, depending on the underground plume around the solid waste landfill disposal site, leachate migration can be detrimental to the environment, human, and wildlife populations located in distant places from the vicinity of the landfill facilities.

(ii) The cost of cleaning up leachate-contaminated environment in most cases is higher than the cost of constructing an engineered landfill.

(iii) The burning of wastes readily contaminates the air and can lead to transboundary pollution with toxic air pollutants, in some instances.

(iv) Environmental and public health consequences associated with unengineered landfills are enormous.

Normally, sanitary landfills are available and possible and are required for the management of the huge solid waste generated. When using sanitary landfills for solid waste management, the following rules are expected to be applied:

(i) Wastes are treated prior to landfilling except for inert wastes.

(ii) Aftercare of closing of landfills is required.

(iii) In sites receiving biodegradable wastes, landfill gas must be collected and used or flared.

(iv) Co-disposal of hazardous wastes with municipal solid wastes is not acceptable.

The most commonly practiced method of solid waste disposal into landfills in many developing and developed countries, where rules guiding solid waste disposal into landfills are not strictly enforced, is the haphazard disposal of all kinds, whether municipal, industrial, or clinical/hospital solid wastes, without segregation, and sometimes along with liquid wastes into landfills. This method is neither hygienic nor safe, however, since this method involves little financial and managerial constraints. In most cases in many countries, these unsanitary landfills (dumpsites) are located in public places surrounded by residential quarters and wetland or other places with seasonally high-water tables (Alimba *et al.*, 2006, 2012; Alimba, 2013).

4. Causes and Types of Landfill Fires

Landfill fires, garbage fires, or waste fires describe fire incidents in which solid waste materials in the landfills are ignited with fire and are burned. Fire ignition of combustible solid wastes at the landfill may be spontaneous and/or accidental (deliberate). Landfill fires pose significant environmental consequences and public health hazards. High risk frequency of landfill fire occurs in non-sanitary landfills and or illegal dump sites than regulated or sanitary landfills, mostly due to accidental or deliberate fire ignition to reduce waste. The frequency of landfill fires also depends on the flammability of the morphological composition of the wastes and the time of the year. Landfill solid wastes containing organic materials, paper and cardboard, plastics and tires, and other rubber materials are more flammable and may experience more frequent landfill fires than landfills containing fewer flammable materials, including metals, glass, and asbestos. Furthermore, fire ignition in most landfills is more frequent during the dry season of the year compared to the wet season (Chavan *et al.*, 2019).

4.1. *Spontaneous combustion/ignition of fire in landfills*

The mountain of solid waste disposed of in landfills receives moisture and air intrusion, which provides oxygen that enhances biological activities, leading to the decomposition of biodegradable components of the solid waste in the landfills. Moisture and air usually intrude into the landfills when it is not covered after the daily waste disposal. Decomposed organic and combustible components of the solid wastes are ignited due to the presence of some gases including methane which enhances combustion. It usually occurs naturally when the heat change resulting from the chemical oxidation of decomposing waste materials increases the temperature of the landfill, leading to fire ignition. This spontaneous combustion, a type of landfill fire, describes the rapid oxidation of decomposable materials in the municipal solid waste facility and it is directly related to the amount of moisture present in the landfill. The bacteria, both aerobic and anaerobic, present in the organic matter of the solid wastes require water to biologically break down organic matter (see Fig. 2 for details).

4.2. *Accidental or deliberate burning of landfill solid wastes*

Open burning covers a wide range of different uncontrolled solid waste combustion practices, including dump fires and pit burning of solid wastes. In many countries, it is a common practice, as it is viewed by many people

$$\underset{\text{(biodegradable organic components)}}{\textbf{Solid wastes}} + H_2O + \underset{(O_2)}{\text{Air}} \xrightarrow[\text{change in temperature}]{\text{Microbes: bacteria}} \textbf{decomposed organic materials} + CH_4 + CO_2 + \textbf{other gases}$$

Fig. 2. Mechanisms of spontaneous fire ignition following generation of combustible gases such as methane and heat during microbial (aerobic) decomposition of organic wastes.

as the cheapest, easiest, and most sanitary means of solid waste disposal by reducing the volume of combustible solid wastes. This type of landfill fire is popular in many countries of the world, most importantly in countries where non-sanitary landfilling or dumping of solid wastes is practiced. For instance, accidental or deliberate landfill or dump burning was recorded in the Deonar and Ghazipur Landfill fire in India, the Cerro Patacon Landfill fire in Panama, and the New Providence Landfill fire in the Bahamas. These dump fires caused numerous devastations in the various communities where they occurred and formed the major headlines of many news outlets. The possible causes of accidental or deliberate landfill fires include the presence of explosive devices in the solid wastes, deliberate ignition of fire by landfill employees or passers-by, and ignition of flammable substances within the source of heat in the landfill. Other possible sources of causes include inattention on the part of landfill employees, improper use of thermal devices, malfunction of electrical installations, work equipment, static electricity, and natural phenomena, including earthquakes and lightning strikes. Waste pickers, also known as scavengers, deliberately set the waste on fire, with the aim of recovering recyclables, including steel wires, among others. Other fires at the landfill occur through the refuse collection vehicles and in the green belt inside the landfill area.

5. Environmental Contamination Due to Landfill Fire

Landfill fires are extremely dangerous to the environment and human health, as well as wildlife health, due to the emission of dangerous chemicals, particulate matter, and smoke from the combustion of various solid waste materials present in the landfill. The consequential effects of waste fire are observed in the various components of the environment. Pollutants emitted when wastes are burnt are discharged into the atmosphere through smoke and particulate matter, while some of the emissions dissolve or are contained in leachates which percolate into the lower ground layers and contaminate groundwater supply, while some contaminate the land surrounding soils and surface water within the vicinity of the landfill.

6. Air Pollution Due to Emissions From Landfill Fire

The effects of landfill fire on the air quality of the surrounding environment have been well studied and reported in numerous scientific and non-scientific publications. Due to landfill fires generally occur at low temperature and under anoxic situation, it is common for dense thick smokes and a variety of harmful toxic gases including dioxins and furans to be released into the air during combustion of wastes containing hydrocarbons, chlorinated materials, and pesticides (Vassiliadou *et al.*, 2009), polynuclear aromatic hydrocarbon (PAH) materials (Stamand *et al.*, 2008), respirable particulate matters ($PM_{2.5}$ and PM_{10}) (Kumar *et al.*, 2015), and heavy metals (Alimba *et al.*, 2006, 2012, 2016). However, among most pollutants released into the air during landfill fire, attention had been particularly concentrated on carbon monoxide, nitrogen oxides, hydrochloric acid vapour, carbon dioxide, hydrogencyanide, methane, volatile organic compounds, persistent organic pollutants, ketones, aldehydes, and PMs (Bihalowicz *et al.*, 2021). Most of the emitted pollutants are highly deleterious to human and wildlife health, while others, including CH_4 and CO_2, are well-known greenhouse gases, which greatly contribute to climate change and cause global warming (Bihalowicz *et al.*, 2021). Figure 3 presents examples of some levels of landfill fires and the possible exposure to smoke, fire inferno, and chemicals present in the smoke and gaseous emissions by residents around the landfills.

Landfill fire emits dense dark smoke, which is capable of spreading and settling down in residential buildings and workplaces within the vicinity of the landfills. Smoke generated from landfill fires may spread obnoxious odors not palatable, and this may last even after successfully extinguishing the fire. Smoke generated from landfill fire may cause haziness, which poses a significant risk to the commuters and passers-by. Furthermore, smoke and odors can have adverse effects on the values of properties: residential and business buildings within the surroundings of landfill sites (Rim-Rukeh, 2014).

Fig. 3. (a) and (b) Aba-Eku landfill in Ibadan, Nigeria; the landfill fire in (a) and (b) was either spontaneously ignited or caused by scavengers who want to reduce volume of waste to be able to pick recyclable metals. (c) Olusosun landfill, Ojota, Lagos, Nigeria. It was not very clear how the fire started, but it destroyed properties and human lives. Also, smokes and chemicals were discharged into the environment.

7. Surrounding Soils and Underground/Surface Water Contamination by Landfill Fire Emissions

Studies have shown that the concentrations of heavy metals and other inorganic and organic compounds present in the ash(es) generated from the aftermath of landfill fire combustion are several times higher than in natural municipal landfill waste. Many of these chemicals are dispersed in the water used to extinguish the landfill fire or in precipitation during rainfall and eventually constitute components of leachates generated from the landfill. The migration of such leachates contaminates the surrounding underground/surface waters and soils. After a landfill fire, a study was conducted to analyze the chemical composition of soils and landfill fire residual samples collected from the landfills. Organochlorophenols (OCP) including hexachlorocyclohexanes (HCH) and DDT, polychlorinated biphenyls (PCBs), polycyclic aromatic hydrocarbons (PAH), and heavy metals and trace elements [cadmium (Cd), mercury (Hg), lead (Pb), tin (Sn), titanium (Ti), phosphorus (P), iron (Fe), manganese (Mn), nickel (Ni), cobalt (Co), arsenic (As), chromium (Cr), copper (Cu), and Mg] were significantly higher compared to soil samples collected from some distance away from the landfill (Øygard *et al.*, 2005, 2014; Raudonyte-Svirbutaviciene *et al.*, 2022; Dabrowska *et al.*, 2023).

8. Exposure Pathways to Landfill Fire Emissions and Health Consequences

It is important to note that without the establishment of exposure, no possible deleterious health effects are expected. Studies have investigated the adverse health effects of exposure to landfill fire. For instance, Lavigne *et al.* (2014) reported that 147 people died during the Leuwigajah dumpsite fire outbreak in Bandung (Indonesia). The 2.7 million cubic meters of waste were on fire and collapsed on the affected people in their settlement. This describes the direct effects of landfill fire on human and wildlife health. Many wildlife and humans living within and around landfills may be trapped in a landfill fire inferno and are killed by the heat of the fire. The most common route of exposure to chemical substances during landfill fire is inhalation of deleterious chemical and particulate matter-laden smoke and gases produced from combusted wastes. Aderemi and Otitoloju (2012) reported that inhalation of toxic smoke and gases emitted from landfill fire caused headache, nausea, and fatigue, also disrupting adequate sleep among exposed residents. Although, the degree of toxicity induced by inhaled substances depends on exposure duration, exposure concentration of the chemicals or particulate matter, and the type of chemical substances. Particulate matter, along with greenhouse gases, may indirectly affect human and wildlife health and the environment by inducing climate change. Many contaminants (chemicals and particulate matter) emitted during landfill fires cause systemic toxicity to the various organs of the body, depending on the route of exposure. For instance, inhalation of such contaminants can cause bronchial asthma, respiratory failure, chronic respiratory illness, cough, congestion, sinus problems, headaches, nose bleeding, upper respiratory junk, chest pains, lung diseases, bronchitis, tuberculosis, and nose/throat irritations (Alimba, 2013). Also, humans can be exposed to contaminants from the landfill fire via dermal contact. Human health effects due to dermal contact include dermal infections (dermatitis), eczema, skin rashes, skin itching, and boils and warts (Alimba, 2013). Contaminants generated during landfill fire are potential carcinogens, genotoxins, systemic toxins, embryotoxins, teratogens, and/or mutagens. They have been shown to elicit myocardial infarction, congenital abnormality, carcinogenesis, developmental abnormality, cardiovascular disease, birth defects, and premature mortality (Alimba, 2013; Ademola *et al.*, 2020). Table 1 presents some gaseous compounds usually generated and discharged into the environment during landfill fire. It also describes the possible effects of these compounds on human, animal, and plant health. Furthermore, their effects on the environment are also presented.

Table 1. Pollutants observed in landfill fire smokes, gases, and leachates, their health, and environmental impacts.

Pollutants	Health Effects	Environmental Effects
Aldehydes	Cause eye and respiratory illness, headaches, and are carcinogens	It can increase toxic load on the environment. It contaminates water and land, which may affect animal health.
Carbon monoxide	Causes dizziness, headaches, and reflexes. Affect mental function, visual acuity, and alertness.	Oxidized to carbon dioxide (which is a greenhouse gas) in the atmosphere.

Table 1. (*Continued*)

Pollutants	**Health Effects**	**Environmental Effects**
Chlorofluorocarbons (CFCs)	Cause dizziness, headaches, and slowed reflexes.	The primary contributors to stratospheric ozone level depletion and are involved in increasing the global warming effect.
Dioxins and furans	They can cause cancer, growth defects, damage DNA, affect immune and reproductive systems.	They increased toxic loading on environment. Cause contamination of water/land and affect animal health. Very toxic and bioaccumulate in the food chain.
Heavy metals	Highly toxic, can bioaccumulate in living system. Non-lethal concentrations cause chronic respiratory or intestinal distress, poisoning of the central nervous system, disruption of the body's hormonal system, inhibition of growth and development in children.	Increase toxic loading on environment, lead to contaminated water and land, and affect animal and plant health.
Hydrochloric acid	Irritation of respiratory tract, causes respiratory illness, and dulls the body's sense organs.	Increased toxicity loading on environment, leads to contaminated water/ land, and can affect animal health.
Hydrogen Sulphide (H_2S)	It is toxic, causes respiratory disease. Healthy people experience shortness of breath, sore throats, breathing difficulties, and irritated eyes.	Contributes to acid rain, may damage vegetation, and causes offensive odors.

(*Continued*)

Table 1. (*Continued*)

Pollutants	Health Effects	Environmental Effects
Ozone (O_3)	Exposure to ozone can injure biological tissues and cells. Reduce lung function, including tightness of the chest, coughing, pain, and breathing difficulty.	Ground-level ozone damages vegetation and ecosystem, and affects animal health.
Nitrogen oxides	Cause respiratory illness, fluid collection in the lungs (oedema), and fibrotic changes.	Contributes to acid rain and ozone formation.
Particulate Matter ($PM_{2.5}/PM_{10}$)	Irritation of respiratory tract, aggravates asthma, and contributes to chronic obstructive pulmonary disease.	Increases toxicity loading on the environment, leads to contaminated water/land, and affects animal health
Polynuclear Aromatic Hydrocarbon (PAHs)	Cancer-causing agent in most animal species, including mammals, fish, and birds.	Increases toxicity loading on environment, leads to contaminated water/ land, and affects animal health.
Volatile Organic Compounds (VOCs)	Direct toxicity, including problems ranging from cancer risks to nervous system disorders. It causes respiratory irritation/illness, affects animal health.	Contributes to low-level ozone (smog), VOCs cause vegetative damage. Leads to contaminated water/ land, affects animal health.
Sulphur Oxides (SO_2)	Increases heart/lung disease, causes acute/chronic respiratory diseases. It causes shortness of breath, sore throats, and breathing difficulties.	It causes vegetative damage, corrodes many materials, and contributes to acid rain.

Source: Aspinwall (2001).

Figure 4 is a schematic presentation of possible human exposure routes to chemicals, particulate matter, and smoke from landfill fire. According to available information on various incidents of landfill fire occurrences, it shows that humans, wildlife, and plants are exposed directly to heat from the fire, which can lead to death and damage to body organs and tissues. Smoke, particulate matter, and gaseous substances generated during burning contain chemicals and particles that may be deleterious to the body. These substances are inhaled through the nostrils into the lungs. These chemicals and particulate matter may induce toxicity on the cells and tissues of the respiratory system and can also be moved into the blood and lymph systems from where they can be transported to the heart, brain, kidney, and other tissues and organs of the body. These chemicals can find their way into leachates, which are capable of contaminating both underground and surface waters, and the soils around the landfill. Humans and wildlife can be exposed to such chemicals by ingesting leachate contaminated underground and surface waters and soils. The chemicals can be absorbed into the body through the skin pores during dermal contact. Ingested and absorbed chemicals through the skin can be moved to the blood and lymphatic systems. The circulatory systems move

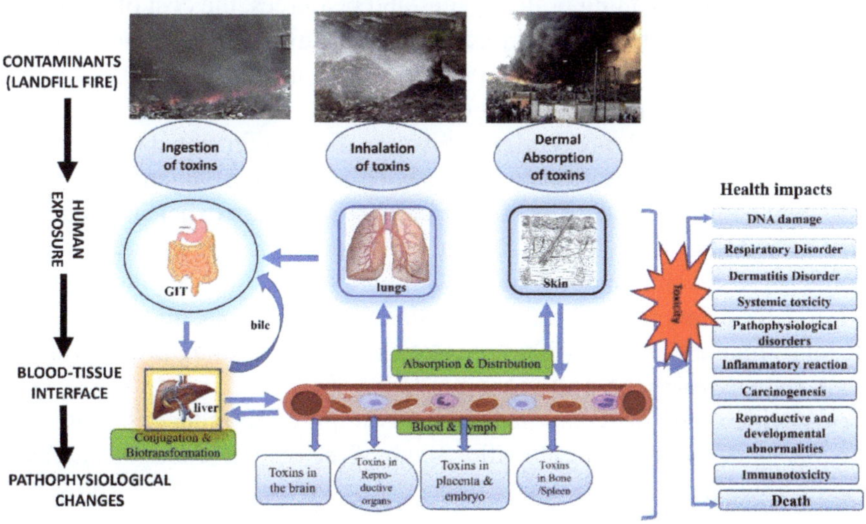

Fig. 4. Systematic presentation of possible emissions from landfill fire, exposure routes of chemicals, particulate matter, and heat emitted from the waste burning, and health effects on living things, including humans.

them to many of the tissues and organs of the body, where they can induce deleterious effects. The chemicals and particulate matter can cause different pathophysiological changes in various organs of the body, which can lead to DNA damage, respiratory disorder, dermatological disorder, systemic toxicity, inflammatory responses, carcinogenesis, reproductive and developmental abnormalities, immunotoxicity, and eventual death of the organisms.

9. Conclusion and Recommendation

Waste fires are an environmental hazard which is currently poorly researched and reported globally. It is capable of occurring in all stages of both public and private solid waste management chains, including during waste collection, transportation, recycling, and disposal at landfills. Landfill fires are capable of causing highly significant damage to human, wildlife, and environmental health through the emission of hazardous chemicals and particulate matter to the atmosphere. Chemicals from landfill fires are capable of contaminating groundwater supplies and the surrounding soil through leachate discharge. The economic costs of waste fires on the various societies of the world are very enormous. This ranges from the cost of cleaning polluted sites and resources, the cost of repairing and rebuilding destroyed structures, to healthcare costs of restoring incapacitating and impaired health induced in human and domesticated animals and wildlife (plants and animals).

Preparation and prevention of waste fires offer the best defence against the risk of injury, death, property damage, environmental degradation, and economic loss. The cost of prevention is less expensive than the cost of fighting waste fires and clean-up costs after a fire has occurred. It is therefore recommended that preventing and taking precaution against waste fires offers the best defence against the risk of sustained injury and death of humans and wildlife, impact on society (including property damage), the environment, and economic loss.

A wide range of recommendations is required to reduce the rate and frequency of landfill fires globally. Some of the recommendations include the following:

(i) practice and maintain engineered/regulated landfilling and incineration of solid wastes, which is capable of detecting and collecting landfill gases to reduce emissions of methane and other gases capable of inducing spontaneous combustion or arson,

(ii) effective waste-site management, including proper storage of combustible wastes,
(iii) effective solid waste transportation from the point of generation to the landfill site is important in preventing waste fires, by sorting and eliminating fire-ignitable materials such as lithium batteries from the collection point,
(iv) quality control site measures, including prohibition of all forms of deliberate burning on the landfill site, thoroughly examining and controlling new wastes to be offloaded in the landfill,
(v) compacting buried refuse to prevent the occurrence of hot spots in the site. Prohibiting smoking on-site and maintaining good security on the site,
(vi) designated staff responsible for landfill fire prevention,
(vii) supply of effective fire-fighting equipment and trained personnel on the site,
(viii) construction of fire breakers around the landfill site to prevent fire spreading across the landfill and to nearby vegetation,
(ix) educating the total population on the calamitous nature of landfill fire. This is expected to enhance good, solid waste disposal and sorting.

Bibliography

Ademola, O.J., Alimba, C.G., and Bakare, A.A. (2020). Reproductive toxicity assessment of Olusosun municipal landfill leachate in *Mus musculus* using abnormal sperm morphology and dominant lethal mutation assays. *Environmental Analysis Health and Toxicology*, 35(2), e2020010. https://doi.org/10.5620/eaht.e2020010.

Aderemi, A. and Otitoloju, A. (2012). An assessment of landfill fires and their potential health effects—A case study of a municipal solid waste landfill in Lagos, Nigeria. *International Journal of Environment and Protection*, 2, 22–26.

Alimba, C.G. (2013). *DNA and Systemic Damage Induced by Landfill Leachates and Health Impacts of Human Exposure to Landfills in Lagos and Ibadan, Nigeria.* Ph.D thesis, Department of Zoology, University of Ibadan, Nigeria, p. 258.

Alimba, C.G. (2024). Alterations in viscera histoarchitecture and organosomatic index as biomarkers of toxicity induced by Aba-Eku and Olusosun solid waste landfill leachates in *Rattus norvegicus*. *Environmental Analysis Health and Toxicology*, 39(2), Article ID: e2024022, 17 pages. https://doi.org/10.5620/eaht.2024022.

Alimba, C.G., Adewumi, O.O., Binuyo, O.M., and Odeigah, P.G.C. (2021). Wild black rats (*Rattus rattus* Linnaeus, 1758) as zoomonitor of genotoxicity and systemic toxicity induced by hazardous emissions from Abule Egba unsanitary landfill, Lagos, Nigeria. *Environmental Science and Pollution Research*, 28, 10603–10621. https://doi.org/10.1007/s11356-020-11325-7.

Alimba, C.G. and Bakare, A.A. (2006). Municipal landfill leachate induced chromosome aberration in rat bone marrow cells. *African Journal of Biotechnology*, 5, 2053–2057.

Alimba, C.G., Bakare, A.A., and Aina, O.O. (2012). Liver and kidney dysfunction in Wistar rats exposed to municipal landfill leachate. *Resources and Environment*, 2, 150–163.

Alimba, C.G., Gandhi, D., Sivanesan, S., Bhanarkar, M.D., Naoghare, P.K., Bakare, A.A., and Krishnamurthi, K. (2016). Chemical characterization of simulated landfill soil leachates from Nigeria and India and their cytotoxicity and DNA damage inductions on three human cell lines. *Chemosphere*, 164, 469–479. https://doi.org/10.1016/j.chemosphere.2016.08.093.

Alimba, C.G., Onajole, A.I., and Bakare, A.A. (2015). Brain dysfunctions in Wistar rats exposed to municipal landfill leachates. *Beni-Suef University Journal of Basic and Applied Sciences*, 4(4), 284–290. https://doi.org/10.1016/j.bjbas.2015.11.003.

Alimba, C.G., Rudrashetti, A.P., Sivanesan, S., and Krishnamurthi, K. (2022a). Landfill soil leachates from Nigeria and India induced DNA damage and alterations in genes associated with apoptosis in Jurkat cell. *Environmental Science and Pollution Research*, 29, 5256–5268. https://doi.org/10.1007/s11356-021-15985-x.

Alimba, C.G., Sivanesan, S., and Krishnamurthi, K. (2022b). Mitochondrial dysfunctions elicited by solid waste leachates provide insights into mechanisms of leachates induced cell death and pathophysiological disorders. *Chemosphere*, 307(Pt 4), 136085. https://doi.org/10.1016/j.chemosphere.2022.136085.

Aspinwall. (2001). Environmental monitoring at Nant-y-Gwyddon landfill site, Cardiff: Environ Aspinwall Report E5169/June/1994/L8:26.

Bihalowicz, J., Rogula-Kozlowska, W., and Krasuski, A. (2021). Contribution of landfill fires to air pollution—An assessment methodology. *Waste Management*, 125, 182–191.

Chavan, D., Lakshmikanthan, P., Mondal, P., Kumar, S., and Kumar, R. (2019). Determination of ignition temperature of municipal solid waste for understanding surface and sub-surface landfill fire. *Waste Management*, 97, 123–130.

Dabrowska, D., Rykala, W., and Nourani, V. (2023). Causes, types and consequences of municipal waste landfill fires—literature review. *Sustainability*, 15, 5713. https://doi.org/10.3390/su15075713.

El-Fadel, M., Findikakis, A.N., and Leckie, J.O. (1997). Environmental impacts of solid waste landfilling. *Journal of Environmental Management*, 50, 1–25.

Kumar, S., Aggarwal, S.G., Gupta, P.K., and Kawamura, K. (2015). Investigation of the tracers for plastic-enriched waste burning aerosols. *Atmospheric Environment*, 108, 49–58. https://doi.org/10.1016/j.atmosenv.2015.02.066.

Lavigne, F., Wassmer, P., Gomez, C., Davies, T.A., Hadmoko, D.S., Iskandarsyah, T.Y.W.M., Gaillard, J.C., Fort, M., Texier, P., Heng, M.B., and Pratomo, I. (2014). The 21 February 2005, catastrophic waste avalanche at Leuwigajah dumpsite, Bandung, Indonesia. *Geoenvironmental Disasters*, 1, 10, 1–12.

Lee, G.F. and Jones, R.A. (1991). Municipal solid waste management: long-term public health and environmental protection. A paper prepared for workshop on landfills and groundwater quality. NWWA national outdoor action conference, Las Vegas, Nevada, May, 15. https://www.gfredlee.com/Landfills/MSWMANAGT.pdf. Accessed 20 December 2024.

Morales, S., Toro, A., Morales, L., and Leiva, G. (2018). Landfill fire and airborne aerosols in a large city: Lessons learned and future needs. *Air Quality, Atmosphere & Health*, 11, 111–121.

Øygard, J., Mage, A., Gjengedal, E., and Svane, T. (2005). Effect of an uncontrolled fire and the subsequent fire fight on the chemical composition of landfill leachate *Waste Management*, 25, 712–718.

Øygard, J.K., Maage, A., and Gjengedal, E. (2014). Estimation of the mass-balance of selected metals in four sanitary landfills in Western Norway, with emphasis on the metal content of the deposited waste and the leachate. *Water Research*, 38, 2851–2858.

Raudonyte-Svirbutaviciene, E., Stakeniene, R., Joksas, K., Valiulis, D., Bycenkiene, S., and Zarkow, A. (2022). Distribution of polycyclic aromatic hydrocarbons and heavy metals in soil following a large tire fire incident: A case study. *Chemosphere*, 286, 131556.

Rim-Rukeh, A. (2014). An assessment of the contribution of municipal solid waste dump sites fire to atmospheric pollution. *Open Journal of Air Pollution*, 3, 53–60.

Stamand, A., Mayer, P., and Blais, J. (2008). Seasonal trends in vegetation and atmospheric concentrations of PAHs and PBDEs near a sanitary landfill. *Atmospheric Environment*, 42, 2948–2958. https://doi.org/10.1016/j.atmosenv.2007.12.050.

Vassiliadou, I., Papadopoulos, A., Costopoulou, D., Vasiliadou, S., Christoforou, S., and Leondiadis, L. (2009). Dioxin contamination after an accidental fire in the municipal landfill of Tagarades, Thessaloniki, Greece. *Chemosphere*, 74, 879–884. https://doi.org/10.1016/j.chemosphere.2008.11.016.

Yaashikaa, P.R., Kumar, P.S., Nhung, T.C., Hemavathy, R.V., Jawahar, M.J., Neshaanthini, J.P., and Rangasamy, G. (2022). A review on landfill system for municipal solid wastes: Insight into leachate, gas emissions, environmental and economic analysis. *Chemosphere*, 309, 136627.

https://doi.org/10.1142/9789819822249_0007

Chapter 7

Military Pollution-Related Metal Transfer Factor for Forest Trees and by Other Plant Species to Food

A. S. Mollah

Department of Nuclear Science and Engineering
Military Institute of Science and Technology
Mirpur Cantonment, Mirpur, Dhaka 1260, Bangladesh

mollahas@gmail.com

Abstract

Heavy metals are released into the environment during military activities through shooting ranges and weapon residues that contain high concentrations of particles comprising depleted uranium (DU), copper (Cu; unleaded), and lead (Pb; leaded ammunition). Due to this, military operations lead to soil pollution with Pb and Cu, as well as other heavy metals, such as Fe, Cr, Ni, and Zn. These metals then move to forest plants and then foods, raising the possibility of human exposition. Due to the deposition of heavy metals from explosions and ammunition, military shooting ranges are recognized sources of environmental contamination. According to a meta-assessment of soil metal levels, raised soil Pb may be regarded as the most serious health threat, whereas a considerable increase in soil Pb, Cu, Fe, Zn, Ni, and Cr content was found in various shooting ranges. Even if the data is conflicting, a growing amount of research has shown that military activities and armed conflicts may be a factor in metal contamination of the environment. In particular, examination of soils from military firing ranges revealed notable Pb, Cu, Ni, Fe, Zn, and Cr pollution. As a result, we review possible sources of metal emissions associated with military activities, and heavy metal exposure pathways and health risks in this work. Transfer factor (TF) is the term used to describe a metal species' capacity to move from the soil into plant roots. The TFs are predicated on the metal's root uptake. Higher TF values (≥1) conceive that the plant is more worthy for phyto-extraction and phyto-remediation and that it absorbs more metal from the soil. Conversely, lower values show that plants respond poorly to metal absorption and that they are fit for human food. The objective of this study is to analyze the heavy metal concentrations in soil and plant samples collected from military gunshot locations. Data analysis, statistical interpretation, and environmental health consequences are all included in the study. A case study on military related heavy elements in two shooting range sites in Dhaka, Bangladesh, was considered for the first time because no such study data are available in the literature. In this chapter, we address the possible metal source emissions related to military activities and the possible pathways for heavy metal and the health exposures associated with them through transfer factors from soil to forest trees and other plant species to food. Different challenges and future directions are also discussed in this study.

Keywords: Heavy metal, concentration, TF, soil, plant, trees, shooting range.

1. Introduction

Shooting ranges used for military training have been found to be significant contributors of heavy metal pollution in terrestrial ecosystems in many studies (Migliorini *et al.*, 2004). Even because shooting ranges only occupy a small area, humans and all other living things may be at risk due to the high levels of heavy metals that have accumulated in their soils and trees (Migliorini *et al.*, 2004; Shoukat *et al.*, 2020; Dinake *et al.*, 2019; Rodríguez-Seijo *et al.*, 2017; Sanderson *et al.*, 2014; Mariussen *et al.*, 2017; Sehube *et al.*, 2017). Due to the buildup of potentially toxic elements (PTEs) in soil that comes from bullets, pollution of shooting range soils is a global concern (Shoukat *et al.*, 2020; Dinake *et al.*, 2019; Rodríguez-Seijo *et al.*, 2017). Bullets containing Pb (94–96.8%) and Zn, Ni, Fe, Cr, and Cu (the remaining percentage) are the cause of contamination in soil shooting ranges. PTEs such as lead, zinc, antimony, arsenic, copper, and nickel make up the majority of the pellets used in ammunition (Shoukat *et al.*, 2020; Dinake *et al.*, 2019; Rodríguez-Seijo *et al.*, 2017). Numerous recent studies have demonstrated that the abrasion of lead bullets as they move through the soil causes a high level of lead pollution in the soils of shooting ranges (Mariussen *et al.*, 2017; Sehube *et al.*, 2017). The second-largest source of lead pollution in the environment, behind the battery sector, is shooting ranges (Barker *et al.*, 2021). Pb exposure in humans can occur through the air, water, soil, and food. It can lead to behavioral issues, low IQ, and even mortality (Lin *et al.*, 1995). Around 400 mg/kg Pb is the essential level for Pb contamination of soils according to USEPA (1996) (Sorvari, 2002) regulatory requirements. These values are frequently surpassed in soils used for shooting ranges, and these sites must be managed to prevent offsite migration or contamination of surface and ground waters. Due to the high levels of lead found in shooting range soils and the potential negative health effects of Pb exposure, the scientific community has paid close attention to the contamination of shooting range soils (Evangelou *et al.*, 2012; Sanderson *et al.*, 2018; Rodriguez-Seijo *et al.*, 2016).

According to Laporte-Saumure *et al.* (2011), lead accounts for 93.1% of the mass of bullets, making it the main component. Characterizing lead at shooting ranges has been the main emphasis of previous research, which frequently only looked at one range (Chrastný *et al.*, 2010; Darling and Thomas, 2007; Martin *et al.*, 2013; Sanderson *et al.*, 2012; Vantelon *et al.*, 2005). Due to variations in soil and other environmental factors, it is frequently challenging to transfer such studies and the management plans that follow to other locations. Co-contaminants caused by Pb bullets in shooting range soils, including as antimony (Sb), copper (Cu), arsenic (As), zinc (Zn), and nickel (Ni), have also been studied more recently (Laporte-Saumure *et al.*, 2011; Robinson *et al.*, 2008; Rodríguez-Seijo *et al.*, 2016; Rooney *et al.*, 1999; Sehube *et al.*, 2017; EPA, 2005; Bannon *et al.*, 2009; Scott, 2001; Hardison *et al.*, 2004; Mariussen *et al.*, 2017; Darling and Thomas, 2007; Cao *et al.*, 2003; Vantelon *et al.*, 2005; USEPA, 2005; Zhu *et al.*, 2024). According to Laporte-Saumure *et al.* (2011), antimony (1.9%) is used to harden lead, copper (4.5%) is utilized in jacketed bullets, and nickel or zinc alloys (0.5%) are used in Cu jacket bullets. Pb may include arsenic as an impurity. Even though these elements only make up a small portion of bullet mass, frequent usage of shooting ranges is likely to cause certain metalloids—particularly Sb and As—to accumulate. Although there aren't many studies on these co-contaminants at ranges, existing research has shown that these metal(loid) s may also be present in these locations at elevated amounts. Bullet residue contamination in shooting range soils is a significant environmental concern globally, primarily due to the accumulation of heavy metals, such as lead (Pb), copper (Cu), antimony (Sb), and mercury (Hg). These contaminants originate from spent ammunition and can pose ecological and human health risks. The most common pollutant found in shooting range soils is lead. Different countries have reported Pb concentrations in shooting range soil:

- **Nigeria:** Average Pb values of 5680 mg/kg were found in the topsoil near the berm of a military shooting range in Ibadan, which is far higher than background levels.
- **China:** Pb values ranging from 1142 mg/kg to 7674 mg/kg were found in a meta-analysis of five shooting ranges, suggesting significant contamination.
- **United States:** Studies showed that up to 41 mg of lead per bullet is abraded during weathering and abrasion, which greatly contributes to soil Pb pollution.

Soils from shooting ranges have been found to contain additional heavy metals besides lead:

- **Antimony (Sb):** High concentrations of Sb, which is used to harden bullets, have been observed in soils, and shooting activities are strongly associated with its distribution.
- **Mercury (Hg):** Long after its use has ended, mercury, which has historically been employed in primers, remains in soils. Its long-term persistence was shown by studies conducted in Switzerland that found mercury amounts above local safety standards.

Contamination of shooting range soils by bullet residue is a common problem with serious health and environmental repercussions. The type of ammunition used, the frequency of shootings, and environmental factors all affect the level of contamination, which varies by region. To reduce these risks, optimal management practices must be implemented and regularly monitored. Due to the weathering and deposition of Pb bullets, lead contamination in soils near shooting ranges poses a serious environmental risk. The amount of this contamination has been the subject of several studies conducted globally, using a variety of approaches to evaluate Pb levels and their effects on the environment. The age of the shooting range, the characteristics of the soil, and management techniques all affect how contaminated the soil is with lead. The highest quantities of lead are frequently found in berm soils, which are constructed to catch bullets. For instance, lead levels in berm soils in Botswana reached 38,406.87 mg/kg, significantly higher than the 400 mg/kg threshold value set by the US Environmental Protection Agency (USEPA). Lead's mobility and bioavailability in soils affect its effects on the environment. Lead behavior is influenced by a number of factors, including the pH of the soil, the amount of organic matter present, and the presence of other minerals. Lead can occasionally seep into groundwater, endangering both the environment and human health. These dangers can be reduced by putting Best Management Practices (BMPs) into practice, such as routine soil monitoring, pH correction, and bullet retrieval. To sum up, lead pollution at shooting ranges is a worldwide issue that necessitates thorough evaluation and management techniques to reduce hazards to the environment and human health. Ecosystems and human health are at risk due to the buildup of heavy metals. Heavy metals can bioaccumulate in soil-dwelling species, which could upset soil ecosystems. Direct contact with polluted soils or ingestion of contaminated food and drink can expose humans. Numerous countries' risk evaluations have suggested possible non-carcinogenic hazards, especially from exposure to lead.

The literature (Fayiga *et al.*, 2016; Evangelou *et al.*, 2012) on the variety of soil-contaminated shooting range locations was methodically reviewed in this study in order to compile information on contamination levels and remediation strategies. The global levels of soil pollution at shooting rage locations were found to be 5–6 mg/kg for As, 3–8 mg/kg for Cu, 6–9 mg/kg for Sb, and 5–8 mg/kg for Pb. Heavy element quantification was done using portable XRF equipment and WDXRF. The amounts of Pb, Sb, Cu, and As pollution in the soil at shooting rage sites were determined to be 5–8 mg/kg, 6–9 mg/kg, 3–8 mg/kg, and 5–6 mg/kg, respectively. These levels were discovered to be greater than the traditional Pb limit of 400 mg/kg. The other elements, such as As, Sb, and Cu, were discovered to be extremely low level. The levels of Pb, Sb, Cu, Zn,

Cd, and As in shooting range soils ranged from 399 to 10,403 mg/kg, 6.57 to 252 mg/kg, 28.7 to 1250 mg/kg, 5.63 to 153 mg/kg, 1.35 to 8.8 mg/kg, and 3.08 to 15.8 mg/kg, respectively. Metalloids were mostly found in the surface soil (0–10 cm) and the stop butt. Firing activity, soil characteristics, climate, and management techniques were all reflected in the contamination distribution. Cu and Cd concentrations were also greater at the berm. With the exception of Ni and Cr, the inter-element correlation coefficient was typically high (0.70–0.93). This suggests that shooting activities had no effect on Ni or Cr. According to principal component analysis (PCA), 74.7% of the soil's total metal load was attributable to metal levels at the berm, target basement, 100 m, and 200 m firing lines. The degree of contamination was determined by corresponding analysis to be Pb > Cd > Cu > Ni > Cr. The area surrounding the berm showed signs of pollution. In order to reduce the current amount of contamination, mitigation measures are necessary.

This study was conducted with the following goals in order to get data on the metal transfer factor for forest trees and other plant species that are related to military pollution to food:

- to conduct a review of the literature on the metal transfer factor for forest trees and other plant species to food that is related to military pollution,
- to evaluate the heavy metals and transfer factor (TF) associated with soil and plant samples from military shooting ranges located in Dhaka, Bangladesh, as a case study because no such information is available in the literature,
- to draw attention to the challenges and potential paths forward regarding heavy metal contamination and soil-to-plant transfer factors associated with military pollution.

2. Literature Review

2.1. *World-wide military-related pollution*

Researchers and scientists are always looking for environmentally appropriate ways to manage and control lead contamination in shooting range soils and trees. For sustainable soil remediation and reclamation initiatives, the financial effects of such techniques have also been a hot topic. Phytoremediation has emerged as a more economical, effective, and ecologically friendly technique for amending heavily contaminated soils in recent years (Mellor and McCartney, 1994; Park and Bae, 2014; Scoriza and Correia, 2019; Rooney *et al.*, 1999; Manninen and Tanskanen, 1993; Dinake *et al.*, 2021; Kabata-Pendias and Pendias, 1992; Tariq and Ashraf, 2016; Evangelou *et al.*, 2012). The United States Environmental Protection Agency (USEPA) has recommended phytoremediation as one of the strategies for managing and controlling lead contamination in shooting range due to its many advantages. The vegetation that grows in and around shooting ranges is also at risk of contamination, in addition to the soils that are located on the range's grounds. Through a variety of chemical and physical processes, plants absorb lead. Since plants absorb this harmful heavy metal, lead poisoning of shooting range soils negatively impacts plants. Since the majority of shooting ranges are unfenced, they serve as grazing grounds for animals and livestock. Pb may migrate across the food chain as a result of this.

Approximately 100,000 military and outdoor shooting ranges are used for shooting sports globally, according to sources (Smith, 2019; Chen *et al.*, 2002). As a result, up to 72,600 tons of lead from munitions are released onto the grounds of these various national ranges each year. In addition, antimony (Sb) is used to harden lead, Ni or Zn are used as alloys in copper jacketed bullets, arsenic (As) coexists with lead and is a component of lead shotgun shells, chromium (Cr) is used as an alloy material for some bullet types, and clay targets contain between 3000 and 40,000 mg of polycyclic aromatic hydrocarbons (PAHs) per kilogram. The preservation of operational preparedness for troops entails military training involving live gunfire, which unavoidably leads to soil and groundwater contamination in the surrounding areas. The Pb–Sb core, which makes up about 95% of the bullet, and the Cu–Zn casing, which makes up about 5% of the bullet, are the main parts of the ammunition used for range training. Pb (>90%), Pb–Sb alloy (<5%), As (<0.5%), and Ni (<0.5%) make up the majority of the core, while a Cu–Zn (90%/10%) alloy makes up the casing. There are firing ranges all over the world now. The careless disposal of ammunition and its fragments during range operations can seriously contaminate range sites, endangering both the natural environment and public health. Land and groundwater resources are essential foundations of sustainable human development. Military activities release toxic metals, such as

- lead (Pb) (from bullets, ammunition),
- cadmium (Cd) (from explosives, batteries),
- arsenic (As) (from chemical weapons, military waste),

Table 1. Typical TF for forest trees.

Tree Species	Metal Accumulation (TF)	Key Metals Accumulated
Birch (*Betula spp.*)	Moderate-High (TF: 0.3-1.5)	Cd, Pb, Zn
Pine (*Pinus spp.*)	Low-Moderate (TF: 0.1-0.5)	Pb, Cu
Oak (*Quercus spp.*)	Low (TF: 0.05-0.3)	Pb, Cr
Poplar (*Populus spp.*)	High (TF: >1)	Zn, Cd

- zinc (Zn) (from coatings, alloys),
- copper (Cu) (from brass casings, wiring),
- nickel (Ni) (from stainless steel, armor),
- uranium (U, DU) (from depleted uranium shells).

These contaminants can persist in the soil and enter the food chain through plants, especially forest trees and other vegetation. The metal transfer factor (TF) is a crucial measure for assessing the uptake of heavy metals from soil into plants, particularly in areas affected by military activities, where pollution from heavy metals, such as Pb, Cd, and As, and others can be significant. When evaluating the uptake of heavy metals from soil to plants, the metal transfer factor (TF) is an essential metric, especially in locations where military pollution is present (such as shooting ranges, explosion sites, and military bases). This component is crucial for assessing the possible dangers of metal pollution in the food chain and forest ecosystems. The transfer factor (TF) is calculated (Mollah and Ferdous, 2025; Mollah *et al.*, 2021; Frissel *et al.*, 2002) as

$$TF = C_{plant}/C_{soil}$$

where C_{plant} is the concentration of metal in plant tissue (mg/kg) and C_{soil} is the concentration of metal in soil (mg/kg).

A TF > 1 indicates high bioaccumulation, meaning the plant accumulates metals efficiently. A typical TF for forest trees is given in Table 1.

Forest trees have low to moderate TFs because they accumulate metals mainly in roots rather than leaves or wood. However, some species show higher uptake:

- Pine (Pinus spp.) → Moderate uptake, stores metals in bark.
- Birch (Betula spp.) → Known for metal accumulation, particularly Zn, Pb, and Cd.
- Oak (Quercus spp.) → Low TF, metals retained in roots.

Table 2. TF for some plants leading to potential food contamination.

Plant Type	TF Range	Metals Absorbed
Grasses (*Poa spp., Festuca spp.*)	0.5 – 2.0	Pb, Cd, Zn
Mosses & Lichens	>1	Pb, U, As
Leafy Vegetables (*Spinach, Lettuce*)	0.3 – 1.8	Cd, Pb, Cu
Cereal Crops (*Wheat, Rice*)	0.1 – 0.8	Cd, As

- Willow (Salix spp.) → High TF, effective for phytoremediation.
- Poplar (Populus spp.) → Moderate TF, used for cleanup of contaminated soils.

Studies show TF values for forest trees:

- Pb: 0.01–0.1 (low).
- Cd: 0.2–0.5 (moderate).
- Zn: 0.3–1.2 (high).
- Cu: 0.1–0.4 (moderate).

The typical TF values for other plants and food crops are shown in Table 2.

- **Grasses and Mosses:** Absorb heavy metals readily, making them good bioindicators.
- **Leafy Vegetables:** Can accumulate heavy metals, posing food safety risks.
- **Cereals:** Generally have lower accumulation but can still absorb dangerous levels in polluted areas.

Grasses, prevalent in many ecosystems, play a significant role in the transfer of heavy metals through the food web. Studies have shown that certain grass species can accumulate heavy metals, which then biomagnify through trophic levels. For instance, research indicates that *Leymus chinensis* exhibits a notable ability to absorb heavy metals from contaminated soils. The study found that metals such as zinc (Zn), arsenic (As), and copper (Cu) tend to biomagnify in organisms at higher trophic levels, suggesting that grasses can serve as entry points for these metals into the food web:

- **High TF:** Grasses (e.g., ryegrass and fescue) have higher TFs than trees, making them important in monitoring contamination.

- **Hyperaccumulators:** Some ferns and shrubs (e.g., *Pteris vittata* for As) accumulate metals efficiently.

Edible Plants (Risk to Food Chain):

- Leafy Vegetables (e.g., lettuce and spinach) → High TF for Cd and Pb.
- Root Vegetables (e.g., carrots and potatoes) → Moderate TF, metals accumulate in roots.
- Fruits (e.g., apples and berries) → Low TF, metals mostly in peels.

Military-affected soils can contaminate food crops through the following:

- Direct uptake from soil (high TF plants).
- Airborne deposition on leaves (e.g., Pb from shooting ranges).
- Water contamination (e.g., heavy metal runoff).
- High-risk species: Leafy greens, root crops, and mushrooms (some fungi hyperaccumulate metals).
- Low-risk species: Tree fruits (apples and pears) and nuts, as metals do not easily transfer to edible parts.

The metal transfer factor (TF) quantifies the efficiency with which plants absorb heavy metals from soil, serving as a critical indicator of potential contamination in the food chain. In regions impacted by military activities, such as shooting ranges and training grounds, soils often exhibit higher levels of heavy metals, including Pb, Cd, and others (Bennett *et al.*, 2007; Suter *et al.*, 2000). Understanding the TF in plants such as grasses and rice is essential for assessing environmental and health risks.

Rice (*Oryza sativa L.*) is a staple food crop, and its contamination with heavy metals poses direct risks to human health. Research conducted in industrial areas of Bangladesh has provided insights into the TF of various metals in rice plants (Evangelou *et al.*, 2012; Chrastný *et al.*, 2010; Malik *et al.*, 2010). The study reported the following average TF values:

- Copper (Cu): 1.21 to 1.28.
- Zinc (Zn): 1.55 to 1.65.
- Cadmium (Cd): 0.086 to 0.095.
- Lead (Pb): 0.0 to 0.038.

- Chromium (Cr): 0.283 to 0.678.
- Cobalt (Co): 0.420 to 0.436.
- Nickel (Ni): 0.0 to 0.003.
- Arsenic (As): 0.01 to 0.022.

These values indicate that rice plants have a higher propensity to accumulate Cu and Zn compared to other metals. The relatively lower TFs for Pb, Ni, and As suggest limited transfer of these heavy metals from soil to rice grains. The uptake of heavy metals in grasses and rice plants has several implications:

- **Food Chain Contamination:** Grasses serve as forage for livestock, and their contamination can lead to biomagnification of metals in animal products consumed by humans.
- **Human Health Risks:** Consumption of rice with elevated levels of heavy metals can pose health risks, including neurological and developmental effects from metals, such as Pb and Cd.
- **Ecosystem Impact:** Heavy metal accumulation can affect soil properties, microbial activity, and general biodiversity.

Given in the following are ways to reduce the risks associated with heavy metal contamination in military-affected areas:

- **Phytoremediation:** Using hyperaccumulators such as *Thlaspi caerulescens* (Zn/Cd) or *Brassica juncea* (Pb). Utilizing plants with high TFs for specific metals to extract contaminants from the soil.
- **Soil amendments:** Adding biochar or lime to reduce metal bioavailability.
- **Crop selection:** Growing low TF crops in contaminated areas.
- **Monitoring:** Regular testing of soil and plants in military-polluted areas.

The use of Pb-containing ordnance might cause Pb to leak into agricultural fields near shooting ranges, endangering the crops and vegetation there. Due to the prolonged residence time of lead in soils caused by its insoluble mineralogical products and the absence of microbial degradation, this toxic heavy metal's bioavailability and bioaccessibility can persist in the soil for an extended period of time. Pb background values in soils typically vary between 10 and 30 mg/kg. Table 3 provides examples

Table 3. Pb uptake by different vegetable species growing in and around shooting sites.

Location and year of study	Total Pb concentration in shooting range soil (mg/kg)	Plant tissue studied	Concentration of Pb in Plant tissue (mg/kg)
Finland (1993)	4,700–54,000	(i) Leaves	(i) 14–70
		(ii) Lingoberries	(ii) 0.3
England (1994)	36–10,620	(i) Roots	(i) 470
		(ii) Stems	(ii) 62
		(iii) Seed pods	(iii) 12
		(iv) Seeds	(iv) 148
New Zealand (1998)	4,000–8,300	(i) Roots	(i) 1347.2–3825.7
		(ii) Leaves	(ii) 9.6–93.7
Switzerland (2001)	44–33,600	(i) Leaves	(i) 0.28–1151.5
USA (2002)	16,200	(i) Beard grass *(polypogon spp.)* and cord grass *(spartina foliosa)*	(i) 2.77–18.1
		(ii) Marsh rosemary *(limonium californicum)*	(ii) 8.96
		(iii) Pickleweed *(s. suberterminalis)*	(iii) 6.25
South Korea (2002)	78.00–165.85	(i) Root	(i) 4.15–6.30
		(ii) Shoot	(ii) 3.77–7.30
USA (2005)	300–4500	(i) Grass tiller	(i) 1390–1450
Switzerland (2008)	14,000–156,000	(i) Leaves	(i) 50–4640
Japan (2008)	19,600	(i) Above-ground plant tissue	(i) 166
Pakistan (2010)	2.0–29.0	(i) Root	(i) 1.0–43.0
		(ii) Shoot	(ii) 15.0–41.0
Switzerland (2011)	500	(i) Root	(i) 50–200
		(ii) Shoot	(ii) 5–18
Switzerland (2012)	466–644	(i) Shoot	(i) 11–62
Finland (2012)	50,000	(i) Leaves	(i) 0.97–30

Table 3. (*Continued*)

Location and year of study	Total Pb concentration in shooting range soil (mg/kg)	Plant tissue studied	Concentration of Pb in Plant tissue (mg/kg)
Switzerland (2013)	500	(i) Shoot	(i) 1.3–5.8
USA (2016)	10,068–70,350	(i) Root (ii) Shoot	(i) 1893–5021 (ii) 252–880
Pakistan (2016)	1,331	(i) Stem	(i) 27.71–82.26
Nigeria (2018)	14.85	(i) Shoot	(i) 12.30
Norway (2018)	47–7189	(i) Grass	(i) 1.3–29
USA (2018)	ND(Not determined)	(i) Root (ii) Shoot	(i) 117 (ii) 81
Switzerland (2018)	471	(i) Shoot	(i) 6

of research conducted over the last 25 years that demonstrate the increasing environmental concern over Pb uptake by plants growing in shooting ranges. Some edible fruits accumulated Pb concentrations of up to 0.3 mg/kg in one of the earliest investigations conducted in Finland, when total Pb concentrations in shooting range soil surpassed background soil values of 240 mg/kg by more than 200 times (Evangelou *et al.*, 2012). Table 4 provides examples of Pb levels in the soils of shooting areas across the different countries (Malik *et al.*, 2010).

Previous research works have demonstrated that Pb is not a necessary element in plants and does not have any nutritional value (Smith, 2019). Nonetheless, it has been shown that this element is naturally present in plants; background quantities of 2.12–2.55 mg/kg have been reported in certain plants. It has been observed that plants are harmful to soil containing 100–500 mg/kg of total Pb. Furthermore, Pb levels in plant tissues between 30 and 300 mg/kg are thought to be hazardous and can have negative consequences, such as lowering plant dry weight, photosynthesis, root growth, and the roots' capacity to absorb water from the soil. Several contamination risk assessment indices and factors, including

Table 4. Pb pollution in the soils of shooting sites in different countries (Smith, 2019).

Location and Year of Study	Number of Studied Shooting Ranges	Concentration of Pb Found in Soil (mg/kg)
Switzerland (2010)	2	621 ± 12.35
Norway (2010)	1	22,000 ± 31.47
USA (2011)	3	20,350 ± 21.62
Canada (2011)	4	24,100 ± 78.54
South Korea (2012)	1	4625 ± 19.03
Australia (2012)	4	6403 ± 00
Finland (2012)	2	30,300 ± 113.48
South Korea (2013)	1	11,800 ± 52.11
Norway (2013)	5	30,000 ± 163.90
China (2014)	1	2563 ± 41.53
Netherland (2014)	1	2158 ± 55.10
Australia (2015)	3	30,600 ± 134.88
Finland (2015)	2	23,000 ± 105.49
China (2015)	3	6019 ± 45.18
Nigeria (2016)	1	17,500 ± 59.74
Australia (2016)	4	2145 ± 21.92
Spain (2016)	1	724 ± 12.47
Pakistan	1	1331 ± 19.04
South Korea (2017)	1	3436 ± 49.13
Norway (2017)	7	2700 ± 25.97
Nigeria (2017)	1	26,933 ± 00
Spain (2017)	1	710 ± 19.16
Switzerland (2017)	2	620 ± 10.41
Botswana (2017)	1	38,300 ± 158.51
Lithuania (2018)	1	654 ± 45.12
Alaska (2019)	1	2800 ± 27.49
Poland (2019)	1	3800 ± 14.03

Table 5. Risk assessment of Pb toward plants (Bennett *et al.*, 2007).

Pollution Risk Index	**$[Pb]_{total}$ in soil (mg/kg)**	**Inference**
Bioconcentration factor (BCF)	500	BCF<1 (low Pb bioavailability)
Translocation factor (TF)	12.30–14.85	TF = 2.91 (efficient translocation of Pb from the root to the shoot)
Translocation factor (TF)	82.36–724.85	TF = 0.11–0.66 (low Pb translocation from roots to shoot)
Biological absorption coefficient (BAC)	82.36–724.85	BAC = 2.06 (the plants showed good Pb phytoextraction)
Bioaccessibility Index	21,900	66% of Pb bioaccessible to plants
Hazard Quotient (HQ)	16,400–27,600	0.109–4.10 (low to high Pb risk to plants)
Bioconcentration factor (BCF)	12,167	BCF = 0.22–1.5 (low to medium Pb bioavailability)
Germination Index (G_{index})	161–10,873	G_{index} = 62–82% (inhibition of germination and plant growth due to Pb)

transfer factor (TF), biological concentration factor (BCF), biological accumulation factor (BAF), hazard quotient (HQ), germination index (GI), root growth inhibition (GI), and bioaccessibility, as indicated in Table 5, can be used to evaluate the toxicity of Pb and, consequently, its pollution risk toward receptors such as plants. A higher than one hazard quotient suggests a potential toxicity risk and its severe impacts, necessitating more plant evaluation (Table 5).

To describe the characteristics of the soil, several writers have employed the SEM-EDX and XRD techniques (Manual of WDXRF, Shimadzu, Japan; Fayiga and Saha, 2016). The results in Figs. 1 and 2 are typical. SEM-EDS revealed a variety of mixed phase particles made of oxidized lead that had silicates of quartz and alumina mixed within them. Shooting range soils gradually convert lead to secondary minerals, and SEM scans revealed different levels of oxidation. Figure 1(a) shows an example of Pb oxides on a particle's surface.

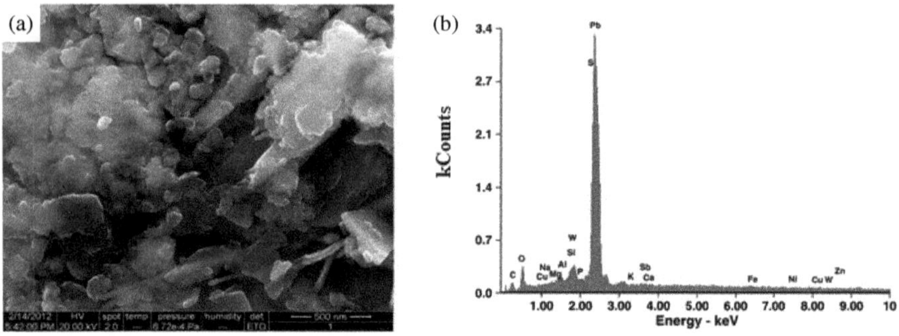

Fig. 1. (a) SEM image of Pb in MB soil. (b) EDS spectra.

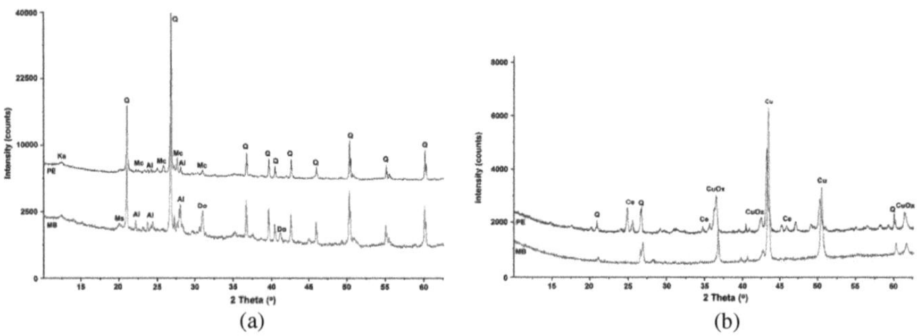

Fig. 2. XRD scans: (a) soil minerals, (b) bullet fragments. Ka = kaolinite Ms = muscovite, Q = quartz, Al = albite, M = microcline, Do = dolomite, Ce = cerussite, Cu = copper, CuOx = copperoxide.

Quartz was the most common mineral in all four soils, according to XRD scans (Fig. 2(a)). XRD analysis of bullet fragments (Fig. 2(b)) revealed copper oxides (CuO) from bullet jackets.

2.2. *Heavy metals in soils in Bangladesh*

The literature in Bangladesh provides sporadic information on heavy metals in vegetables and soil (Majumder *et al.*, 2021; Mizan *et al.*, 2023; Mallick *et al.*, 2019; Uddin *et al.*, 2016). However, there is no information on the plants and soil in the vicinity of the military shooting ranges. Alam *et al.* carried out a Critical Review of Lead Pollution in Bangladesh in 2021 (Mizan *et al.*, 2023). The reviewed research found 175 Pb-contaminated sites in Bangladesh. Pb concentrations in the air (0.092–376.68 $\mu g/m^3$, mean 20.91 $\mu g/m^3$), river water (0.092–18.6 mg/l, mean 1.06 mg/l), river sediments (4.93–69.71 mg/kg, mean 31.97 mg/kg), fish (0.16–30.86 mg/kg, mean 5.08 mg/kg), soil (7.33–445.67 mg/kg, mean 90.32 mg/kg), vegetables (0.24–23.11 mg/kg, mean 4.32 mg/kg), and diet items (0.021–413.92 mg/kg, mean 43.23 mg/kg) were found to exceed relevant guidelines set forth by the FAO, the United States Environmental Protection Agency (USEPA), and the Bangladesh Standard Testing Institution (BSTI) (USEPA, 2005; ITRC, 2003; NIOSH, 2009; WHO, 1996; Ministry of Environmental Protection of the People's Republic of China, 2018).

There is no study on heavy metal pollution in the shooting range soil in Bangladesh. For the first time, a study has been undertaken to determine the heavy metals' contribution in shooting range soil to compare with the reported world values in shooting range soil. According to studies, soils and plants close to industrial zones in Bangladesh contain higher concentrations of heavy metals, such as Pb, Cd, Cr, Ni, Cu, and As. Concentrations of Cr, Ni, Cu, As, Cd, and Pb, for example, ranged from 0.96 to 14.04 mg/kg, 0.71 to 18.39 mg/kg, 1.02 to 34.44 mg/kg, 1.2 to 11.21 mg/kg, 0.44 to 3.31 mg/kg, and 2.01 to 28.86 mg/kg, respectively, according to study conducted in the Tangail district. These levels could be harmful to the environment and human health. Tables 5 and 6 summarize the literature values on heavy metal concentrations in soil and vegetable samples in the Savar, Dhaka region (Mallick *et al.*, 2019). Table 6 makes it clear that the amounts of Cu, Zn, Cd, Ni, and Pb (apart from Cr values) in the soil under investigation were lower than those documented by international organizations. Table 7 makes it clear that the concentrations of Cu, Zn, Cd, Ni, and Pb (with the exception of Ni values) in the vegetable under investigation were higher than those reported by international organizations. This suggests that the vegetables had a larger transfer factor.

Table 6. Mean heavy metal concentration (mg/kg) of collected soil in different zones of Savar and its nearby areas.

Regions	Cr	Pb	Cu	Zn	Ni	Cd
Zone 1	2620	7.25	26.2	205	18.9	3.35
Zone 2	54.2	4.22	25.5	243	26.0	2.23
Zone 3	10200	6.11	18.5	247	9.21	5.11
Zone 4	50.1	41.5	58.7	286	16.1	0.87
Mean of four zones	3220	14.8	32.2	245	17.6	2.89
Background value	19.7	90.7	75.7	38.0	35.3	1.56
FAO/WHO limit	100	100	100	300	50	3
China limit	200	120	100	250	100	0.60

Table 7. Mean heavy metal concentration (mg/kg) of collected vegetable samples in Savar and nearby area.

Regions	Cr	Pb	Cu	Zn	Ni	Cd
Zone 1	1.06	1.61	4.15	98.1	0.30	0.92
Zone 2	2.09	1.65	4.59	118	0.25	1.07
Zone 3	10.7	0.45	2.79	58.3	0.35	1.21
Zone 4	1.20	1.45	4.20	64.0	0.20	0.58
Mean of four zones	3.76	1.29	3.93	84.6	0.28	0.95
FAO/WHO limit	2.3	0.1	40	9.4	10	0.05

Table 8. Typical TF for heavy metal in Bangladesh.

Heavy Metal	Transfer Factor (TF)
Copper (Cu)	0.786
Cadmium (Cd)	0.759
Zinc (Zn)	0.569
Lead (Pb)	0.432
Manganese (Mn)	0.167
Iron (Fe)	0.050

2.3. *Typical TF value in Bangladesh*

Heavy metal pollution of plants and soils is a major issue in Bangladesh, especially in industrial areas. Although there aren't many studies specifically on pollution from military sources, studies conducted in industrial areas offer valuable information about the movement of heavy metals from soil to plants, which can guide mitigation plans for different types of contamination (Table 8).

This suggests that vegetables in the Jashore (Bangladesh) area have higher uptake of Cu, Cd, and Zn compared to Mn and Fe.

3. Heavy Metals in Shooting Range Soils in Bangladesh: A Case Study

The historical deposition of lead-based ammunition is the main cause of the high Pb values in the soils of the shooting range. Localized hotspots are suggested by the large variability (high SD) in Pb concentrations, which is probably caused by particular shooting activity or areas where bullets accumulate. The more even distribution of zinc and nickel, on the other hand, suggests that these metals may come from more dispersed sources or have distinct mobility properties in the soil. For the first time, the heavy metals in shooting range soil and plants in two locations were analyzed.

3.1. *Sample collection and preparation for heavy metal analysis*

In order to obtain representative and trustworthy results, accurate measurements of heavy metals in soil necessitate the use of appropriate collection and preparation methods. A number of soil and plant samples were taken from two shooting range locations and properly identified in a plastic container to prevent cross-contamination. Two shooting range sites are Mirpur Cantonment (Mirpur 23° 49' 20.4600" N and 90° 21' 55.5012" E) and Dhaka Cantonment (Altitude: 23° 49' 19.92" N; Longitude: 90° 24' 29.88" E) for soil and plant sample collection. It is significant to remember that none of the studies on the potential pollution danger from heavy metal deposition in these shooting ranges have been conducted, despite the fact that all of them have been in use for the past 35 years. A schematic diagram of sample collection is shown in Fig. 3. To create three composite samples that are typical of the upper, middle, and bottom berm soils, four samples were also taken from each of the three sampling locations within the sampling area (Fig. 3). In each of the two shooting ranges, berm soil samples were taken with a soil recovery probe down to a depth of 20 cm. This soil is a representation of the deeper soil layer (10–20 cm), the subsurface layer (5–10 cm), and the surface layer (0–5 cm). As a barrier for bullets and shoots after they have struck the target, the berm collects the greatest percentage of shots and bullets. Three sections made up the shooting range's berm area: berm-upper, berm-middle, and berm-bottom. Since it covers a wide region and ensures that the sample accurately represents the study site, this sampling strategy was used. The five shooting ranges were 200 meters distant from the control soil samples, which were used as background samples. Before being sent to the Military Institute of Science and Technology for additional sample processing and analysis, the gathered samples were first placed in plastic containers.

A conventional approach was followed to prepare three sub-samples for statistical analysis. To prevent cross-contamination, a control soil and plant sample was taken for comparison, a long way from the shooting range location, and properly identified in a clean plastic container. For

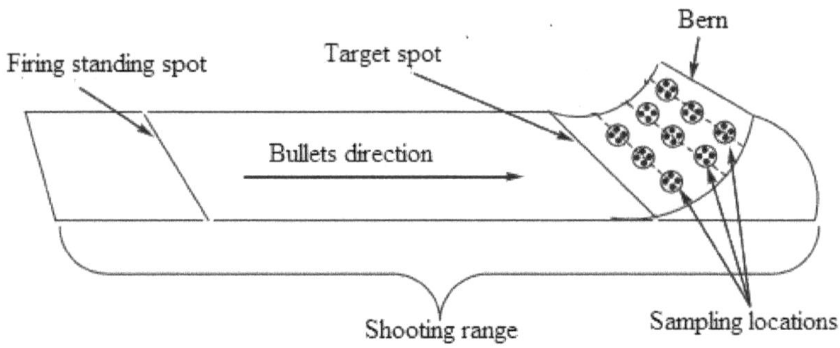

Fig. 3. Diagram of the sampling locations within the shooting range's berm.

analysis, soil samples were allowed to air dry for a few days at room temperature (25–30°C) in a dust-free environment. Using a mortar and pestle, all samples were crushed, and the dried sample was ground. The sample was passed through a 2 mm sieve to obtain homogeneous. Plant samples were dried at 80 °C in an oven and crushed with a mortar and pestle. To guarantee a consistent distribution of heavy metals, every sieved soil sample was thoroughly mixed. Similarly, to achieve a consistent distribution of heavy metals, all oven-dried plant samples were well mixed (Chrastný *et al.*, 2010; Rodríguez-Seijo, 2016; Bannon *et al.*, 2009; Lawal *et al.*, 2019; Cao *et al.*, 2003; Bandara and Vithanage, 2016).

3.2. *Analytical techniques*

Assessing the environmental and health concerns of shooting range soils requires knowing the quantities of heavy metals present. Samples of plants and soil were gathered from various areas of the shooting range. Portable XRF and WDXRF for non-destructive methods were used to measure the amounts of heavy metals. A standard sample supplied by the equipment manufacturer was used to calibrate the portable XRF and WDXRF XRF-1800 (Wavelength Dispersive X-ray Fluorescence (WDXRF)) from Shimadzu Instruments as a non-destructive analytical technique (Manual of WDXRF, Shimadzu, Japan; Fayiga and Saha, 2016). Figure 4 displays a schematic diagram of the apparatus. A quick and non-destructive analytical method for determining the presence of heavy metals in soil is X-ray fluorescence (XRF). It eliminates the need for intricate sample digestion processes and enables elemental analysis *in situ* or in a lab. High-energy X-rays are used to bombard a sample for the XRF to function. The X-ray

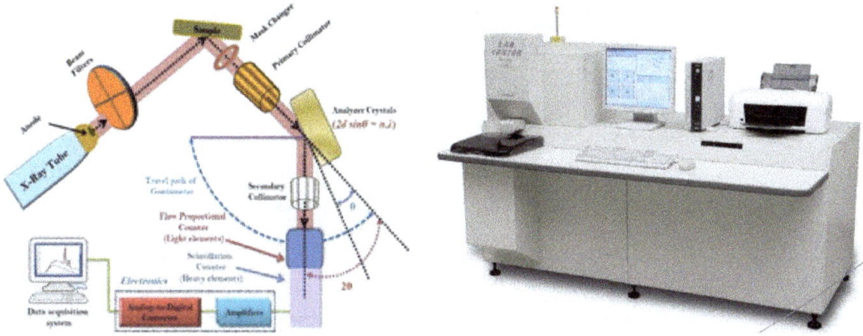

(a) (b)

Fig. 4. (a) Schematic diagram for working principle of WDXRF; (b) photograph of WDXRF equipment.

energy is absorbed by the sample's atoms, which then release secondary (fluorescent) X-rays. Heavy metal identification and quantification are made possible by the characteristic X-rays that are released by particular elements. To ensure accuracy, replicate measurements Certified Reference Materials (CRM) and blank tests were conducted.

4. Results and Discussion

4.1. *Heavy metal contamination*

The results of available heavy metals in shooting range soils are shown in Table 9. Average concentrations of Pb, Zn, Cu, Ni, Cr, and Fe are found to be 420160.75 $\pm$ 13.27 mg/kg, 2708.75 $\pm$ 0.0554 mg/kg, 17845 $\pm$ 0.5056 mg/kg, 310.5 $\pm$ 0.0120 mg/kg, 467.75 $\pm$ 0.0322, and 467.75 $\pm$ 0.0322 mg/kg, respectively, for location A (Mirpur cantonment). Average concentrations of Pb, Zn, Cu, Ni, Cr, and Fe are found to be 97439.25 $\pm$ 4.187 mg/kg, 2628.9 $\pm$ 0.1699 mg/kg, 10965.1 $\pm$ 0.4481 mg/kg, 612.93 $\pm$ 0.0222 mg/kg, 804.25 $\pm$ 0.0184 mg/kg, and 256180 $\pm$ 7.2989 mg/kg, respectively, for location B (Dhaka cantonment). The results of the control soil sample far away from the shooting range are also shown in Table 9 for comparison. Average concentrations of control soil samples of Pb, Zn, Cu, Ni, Cr, and Fe are found to be 1627 $\pm$ 0.0321 mg/kg, 3351.75 $\pm$ 0.1783 mg/kg, 680.25 $\pm$ 0.0057 mg/kg, 725.25 $\pm$ 0.0107 mg/kg, 888.75 $\pm$ 0.0138 mg/kg, and 273808.25 $\pm$ 0.9196 mg/kg, respectively. For a 68% confidence level, the standard deviation (1σ) is also shown in Table 9. The concentrations

Table 9. Heavy metal concentration in shooting range soils.

Elements	Location A (mg/kg)	Location B (mg/kg)	Control (mg/kg)
Pb	420160.75 + 13.27	97439.25 + 4.187	1627 + 0.0321
Zn	2708.75 + 0.0554	2628.9 + 0.1699	3351.75 + 0.1783
Cu	17845 + 0.5056	10965.1 + 0.4481	680.25 + 0.0057
Ni	3010.5 + 0.0120	2612.93 + 0.0252	725.25 + 0.0107
Cr	1467.75 + 0.0322	1804.25 + 0.0184	888.75 + 0.0138
Fe	251207.5 + 8.0322	256180 + 7.2989	273808.25 + 8.9196

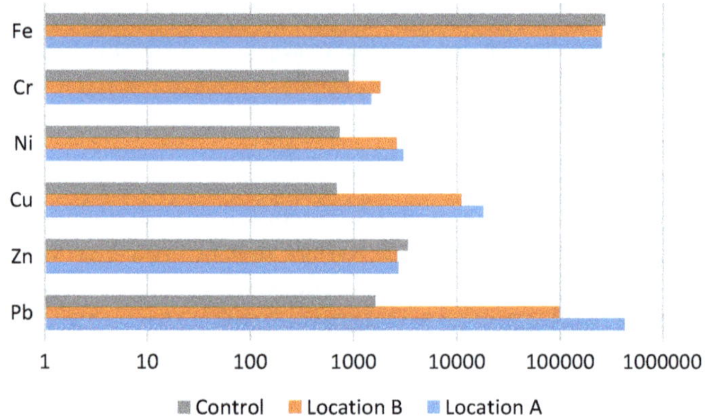

Fig. 5. Comparison of heavy metals for two locations with control sample.

of Pb, Cu, Ni, Cr, and Fe are found to be higher in both locations in shooting range soil samples (except Zn) in comparison to the control sample (background). This means that shooting range soils are highly contaminated with Pb, Cu, Ni, Cr, and Fe except Zn. According to the corresponding analysis, the degree of contamination was Pb >Fe> Cu > Ni > Cr for both the locations. Mitigation steps are necessary to reduce the contamination level as it stands. Comparison results are shown in Fig. 5. The statistical correlation co-efficients are shown in Table 10.

The heavy metal concentrations in plants and TF are shown in Table 11. The results indicate that grass and forest plants accumulate heavy metals differently, with grasses showing slightly higher TF values. The translocation of metals varied, with Cu demonstrating the highest TF, suggesting easier mobility within plants. This finding aligns with previous research indicating Cu's affinity for translocation. Pb and Zn, being less mobile, showed lower TF values, implying their strong retention in roots.

Table 10. Correlation coefficients between total metal content in shooting range soils for location A.

Variables	Pb	Zn	Cu	Ni	Fe	Cr
Pb	1					
Zn	0.627981	1				
Cu	−0.14301	0.679306	1			
Ni	−0.57702	0.266476	0.887418	1		
Fe	0.05224	0.807421	0.980534	0.783747	1	
Cr	0.608416	0.989238	0.691786	0.296637	0.820103	1

Table 11. Heavy metal concentrations in plants and TF.

Plant Type	Pb (Root)	Pb (Shoot)	Zn (Root)	Zn (Shoot)	Cu (Root)	Cu (Shoot)	TF (Pb)	TF (Zn)	TF (Cu)
Grass 1	120	80	75	50	40	30	0.67	0.67	0.75
Grass 2	140	100	85	60	45	35	0.71	0.71	0.78
Forest 1	180	110	90	65	50	38	0.61	0.72	0.76
Forest 2	160	105	80	55	42	33	0.66	0.69	0.79

5. Challenges and Future Direction

Due to the complicated climatic conditions, varying plant responses, and the different forms of contamination, military pollution-related metal transfer factors for forest trees and other plant species provide considerable issues. Assessing the dangers to ecosystems and human health requires an understanding of how metals travel from polluted soil to plants and then through the food chain. Future studies should concentrate on site-specific transfer variables, climate model incorporation, plant and animal bioaccumulation, and the development of standardized techniques. We can better anticipate and control the effects of military-related contamination on food sources and forest ecosystems by tackling these issues, which will eventually result in better environmental management and repair techniques. There are a number of difficulties in identifying heavy metals in soil samples from military shooting ranges, and future research in this area aims to increase the precision, sensitivity, and effectiveness of detection techniques. The following are some of the primary obstacles and possible paths forward:

5.1. *Challenges with metal transfer factors for forest trees and other plant species to food associated with military pollution*

- Chemical agents, explosives, and metals from ammunition are just a few of the contaminants that military ranges can release into the environment. Understanding how metals move through soil, plants, and the food chain is made more difficult by the mixing of metals (such as lead, copper, zinc, and cadmium) in various chemical forms.
- Military ranges and the surrounding soil are frequently diverse, with differing degrees of moisture, organic matter, pH, and pollution. It is challenging to determine consistent transfer factors for various forest trees and plant species because of these characteristics, which affect metal mobility and bioavailability.
- The capacity of various plant species, particularly forest trees, to absorb and accumulate heavy metals from contaminated soil varies. The root systems of different plant species, their ability to withstand metal stress, and the physiological mechanisms behind metal uptake, transport, and storage all influence this heterogeneity.
- Soil characteristics, such as pH, organic matter, and soil texture, have a significant impact on the bioavailability of heavy metals to plants. Metals can be found in soil in forms that plants can easily absorb or in forms that are bound and not bioavailable. As a result, evaluating transfer factors can be difficult when estimating the amount of metal that plants can absorb.
- Since metals may be deposited at different soil depths, the deep and widespread root systems of forest trees make it more difficult to evaluate metal uptake. Additionally, changing climatic conditions (such as excessive rainfall or disturbances) may cause metals to be present at different soil depths, which could impact trees' ability to absorb metals.
- After being absorbed by plants, metals are moved to edible components (such as leaves, fruits, and seeds) depending on a number of variables, such as the metabolic processes of the plant and the surrounding environment. An ongoing difficulty is figuring out how and to what degree metals go to plant components that humans or wildlife can eat.
- Plant development and metal uptake are influenced by climate variables, such as temperature, precipitation, and seasonal variations. Predicting metal levels in plants over time may become more difficult

due to these variables, which may have an impact on the metal transfer process and change from year to year.

- Metal uptake and transmission may be impacted by interactions among many plant species, microbes, and other organisms in the forest ecosystem. For instance, soil fauna may change the availability of metals to plants, and mycorrhizal fungi may affect the transmission of metals to trees.
- Metal transfer factors (TFs) are usually computed for particular ecosystems, but there aren't many standardized models for military pollution environments, so there aren't many trustworthy TF values to direct risk assessments or cleanup.
- One major worry is the movement of heavy metals from plants to humans and animals. Serious health hazards can arise from the bioaccumulation and biomagnification of metals caused by the consumption of forest trees and other plants contaminated by military pollution by humans or animals.

5.2. *Prospects for metal transfer factors associated with military pollution*

- To ascertain the transfer factors for different metal pollutants in military ranges, site-specific research should be carried out. To create precise models, this will need to take into account regional soil characteristics, flora kinds, and the particular metals found in the range area.
- To learn how plants might be utilized to lessen heavy metal contamination in military areas, phytoremediation approaches could be researched. It is possible to ascertain a plant species' potential contribution to lowering metal mobility and transfer by identifying those with high metal uptake or tolerance.
- The flow of metals from the soil to the roots and then through plants should be the main focus of research. Examining root-to-shoot translocation, the function of mycorrhizal fungi in metal uptake, and the effects of various soil types (such as sandy, clay, and loamy) on metal transfer to plants are all part of this.
- A more thorough knowledge of how shifting environmental conditions impact metal transfer will be possible with the inclusion of climate models that forecast temperature, precipitation, and seasonal

variations. This can assist in forecasting possible hazards under various climatic scenarios in the future.

- Species that are more vulnerable to contamination may be found through specific studies on the capacity of forest tree species to acquire metals from military pollution. The possibility that these species will collect metals in edible sections, raising the risk of transfer to humans or wildlife, should also be evaluated in studies.
- How metals move up the food chain from plants to herbivores and predators requires more research. Assessing the overall effects of contamination on ecosystems and human food supplies will be made easier by researching metal uptake by various trophic levels.
- To standardize transfer factor values and create best practices for risk assessment, military, environmental, and academic institutions must work together. This could contribute to the development of general models for evaluating pollution at military ranges.
- Creating standards for the cleanup of heavy metal contamination at military installations should be another area of future study. Finding plant species that can lessen metal pollution or incorporating techniques such as soil amendments to stabilize metals could be part of this.
- To properly evaluate the environmental dangers related to heavy metal contamination at military shooting ranges, future research should concentrate on developing standardized procedures, encouraging on-site study, and integrating real-time monitoring.
- On-site, real-time monitoring with increased sensitivity for detecting trace amounts of heavy metals may be made possible by new technologies, such as electrochemical sensors and portable X-ray fluorescence (XRF).
- To increase the sensitivity and selectivity of heavy metal detection in soil samples, sensors based on nanomaterials and nanotechnology may be used. Faster and more economical analysis may be made possible by these technologies.
- Interpreting complicated environmental data and spotting trends and patterns in contamination over time may be made easier with the application of artificial intelligence and machine learning. Potential sources and contamination hotspots can be predicted using AI systems.
- Effective risk management and consistency of results will be aided by the development of common worldwide standards and regulations for the assessment of contamination at military ranges.

6. Conclusion

A systematic evaluation of the heavy metal pollution in military shooting ranges across several nations has been conducted. In certain instances, it is discovered that the levels of heavy metal contamination in soil and forest plants exceed the international thresholds. This suggests that the distribution of heavy metal contamination in soil and plants throughout the shooting ranges is not uniform. There is no information on the soil and vegetation of military shooting ranges in Bangladesh. Nonetheless, this work reviews and reports some data on heavy metal contamination in plants and soil for ready reference. In certain instances, plant samples from various parts of Bangladesh had a slightly greater level of heavy metal pollution. The level of heavy metal pollution in soil and plant samples from two different locations within the military shooting range site was assessed for the first time. According to the preliminary study, the Pb concentration in the military shooting ranges was three to four times higher than the international limit, but the results were equivalent to those for other nations. Due to the deposition of heavy metals from ammunition, shooting ranges are a contributing factor to environmental contamination. These metals can be absorbed by plants and build up in the soil, endangering both human health and ecosystems. This study uses the TF to assess plants' capacity to transfer metals from roots to shoots and examines the concentrations of heavy metals in shooting range vegetation. One major ecological worry is the poisoning of forest and grass species in shooting range areas by heavy metals. In order to understand plant uptake efficiency, this study will measure the amounts of heavy metals such as Pb, Cu, Zn, Fe, Ni, and Cr in soil and plant samples taken from a shooting range. Additionally, TF values were assessed for limited plants. To evaluate differences in metal accumulation between plant species and environmental factors, statistical analysis was done.

The heavy metals in the current preliminary investigation were quantified using portable XRF equipment and WDXRF. Other methods, including AAS and NAA, will be employed in the future for comparison. In order to create baseline data for quick reference and mitigation techniques in the shooting range areas, more research is required. This study demonstrates that shooting range environments contribute to heavy metal accumulation in vegetation, with varying translocation efficiencies. Understanding these patterns can inform environmental management strategies and phytoremediation efforts to mitigate contamination impacts. In this chapter, the difficulties and future direction are also covered.

Bibliography

Al Mizan, Mamun, M.A.H., and Islam, M.S. (2023). Metal contamination in soil and vegetables around Savar tannery area, Dhaka, Bangladesh: A preliminary study for risk assessment. *Heliyon*, 9(3), e13856. https://doi.org/10.1016/j.heliyon.2023.e13856.

Bandara, T. and Vithanage, M. (2016). Phytoremediation of shooting range soils. In: Ansari, A.A., Gill, S.S., Gill, R., *et al.* (Eds.), *Phytoremediation*. Springer International Publishing, Cham, pp. 469–488.

Bannon, D.I., Drexler, J.W., Fent, G.M., *et al.* (2009). Evaluation of small arms range soils for metal contamination and lead bioavailability. *Environmental Science & Technology*, 43, 9071–9076.

Barker, A.J., Clausen, J.L., Douglas, T.A., Bednar, A.J., Griggs, C.S., and Martin, W.A. (2021). Environmental impact of metals resulting from military training activities: A review. *Chemosphere*, 265, 129138.

Bennett, J.R., Kaufman, C.A., Koch, I., *et al.* (2007). Ecological risk assessment of lead contamination at rifle and pistol ranges using techniques to account for site characteristics. *Science of the Total Environment*, 374, 91–101.

Cao, X., Ma, L.Q., Chen, M., Hardison, D.W., and Harris, W.G. (2003). Lead transformation and distribution in the soils of shooting ranges in Florida, USA. *Science of the Total Environment*, 307, 179–189. https://doi.org/10.1016/S0048-9697(02)00543-0.

Chen, M., Daroub, S.H., Ma, L.Q., Harris, W.G., and Cao, X. (2002). Characterization of lead in soils of a rifle/pistol shooting range in central Florida, USA. *Soil and Sediment Contamination*, 11, 1–17. https://doi.org/10.1080/20025891106664.

Chrastný, V., Komárek, M., and Hájek, T. (2010). Lead contamination of an agricultural soil in the vicinity of a shooting range. *Environmental Monitoring and Assessment*, 162, 37–46. https://doi.org/10.1007/s10661-009-0774-3.

Darling, C.T.R. and Thomas, V.G. (2007). The distribution of outdoor shooting ranges in Ontario and the potential for lead pollution of soil and water. *Science of the Total Environment*, 313, 235–243. https://doi.org/10.1016/S0048-9697(03)00328-0.

Dinake, P., Kelebemang, R., and Sehube, N.J.S. (2019). A comprehensive approach to speciation of lead and its contamination of firing range soils: A review. *Soil and Sediment Contamination*, 28, 431–459. https://doi.org/10.1080/15320383.2019.1597831.

Dinake, P., Mokgosib, S.M., Kelebemanga, R., Kereeditsea, T.T., and Motswetla, O. (2021). Pollution risk from Pb towards vegetation growing in and around shooting ranges – a review. *Environmental Pollutants and Bioavailability*, 33(1), 88–103. https://doi.org/10.1080/26395940.2021.1920467.

EPA (Environmental Protection Agency). (2005). Best management practices for lead at outdoor shooting ranges. EPA-901-B-01-001. Available at: http://www.epa.gov/region2/waste/leadshot.

Evangelou, M.W.H., Hockmann, K., Pokharel, R., Jakob, A., and Schulin, R. (2012). Accumulation of Sb, Pb, Cu, Zn and Cd by various plants species on two different relocated military shooting range soils. *Journal of Environmental Management*, 108, 102–107.

Fayiga, A.O. and Saha, U.K. (2016). Soil pollution at outdoor shooting ranges: Health effects, bioavailability and best management practices. *Environmental Pollution*, 216, 135–145. https://doi.org/10.1016/j.envpol.2016.05.062.

Frissel, M.J., Deb, D.L., Fathony, M., Lin, Y.M., Mollah, A.S., Ngo, N.T., Othman, I., Robison, W.L., Skarlou-lexiou, V., Topcuoğlu, S., Twining, J.R., Uchida, S., and Wasserman, M.A. (2002). Generic values for soil-to-plant transfer factors of radiocesium. *Journal of Environmental Radioactivity*, 58(2–3), 113–128. https://doi.org/10.1016/s0265-931x(01)00061-3.

Hardison, D.W., Ma, L.Q., Luongo, T., *et al.* (2004). Lead contamination in shooting range soils from abrasion of lead bullets and subsequent weathering. *Science of the Total Environment*, 328, 175–183.

ITRC (Interstate Technology and Regulatory Council). (2003). Characterization and remediation of soils at closed small arms firing ranges. Available at: https://itrcweb.org/home.

Kabata-Pendias, A. and Pendias, H. (1992). *Trace Elements in Soils and Plants*, 2nd Edition. CRC Press, Boca Raton, Florida, p. 365.

Laporte-Saumure, M., Martel, R., and Mercier, G. (2011). Characterization and metal availability of copper, lead, antimony and zinc contamination at four Canadian small arms firing ranges. *Environmental Technology*, 32(7), 767–781. https://doi.org/10.1080/09593330.2010.512298.

Lawal, I.A., Uba, S., and Ekwumemgbo, P.A. (2019). Concentrations of lead, cadmium and copper in soil samples at a shooting range in Nasinta, Katsina, Nigeria. *Nigerian Research Journal of Chemical Sciences*, 7(2), 133–142.

Lin, Z., Comet, B., Qvarfort, U., and Herbert, R. (1995). The chemical and mineralogical behaviour of Pb in shooting range soils from central Sweden. *Environmental Pollution*, 89, 303–309.

Majumder, A.K., Al Nayeem, A., Islam, M., Akter, M.M., and Carter, W.S. (2021). Critical review of lead pollution in Bangladesh. *Journal of Health & Pollution*, 11(31).

Malik, R.N., Husain, S.Z., and Nazir, I. (2010). Heavy metal contamination and accumulation in soil and wild plant species from industrial area of Islamabad, Pakistan. *Pakistan Journal of Botany*, 42, 291–301.

Mallick, S.R., Proshad, R., Islam, M.S., Sayeed, A., Uddin, M., Gao, J., and Zhang, D. (2019). Heavy metals toxicity of surface soils near industrial vicinity: A study on soil contamination in Bangladesh. *Archives of Agriculture and Environmental Science*, 8(1), 356–368. https://doi.org/10.26832/24566632.2019.040401.

Manninen, S. and Tanskanen, N. (1993). Transfer of lead from shotgun pellets to humus and three plant species in a Finnish shooting range. *Archives of Environmental Contamination and Toxicology*, 24, 410–414.

Manual of WDXRF, Shimadzu, Japan.

Mariussen, E., Heier, L.S., Teien, H.C., Pettersen, M.N., Holth, T.F., Salbu, B., and Rosseland, B.O. (2017). Accumulation of lead (Pb) in brown trout (Salmo trutta) from a lake downstream a former shooting range. *Ecotoxicology and Environmental Safety*, 135, 327–336. https://doi.org/10.1016/j.ecoenv.2016.10.008.

Mariussen, E., Johnsen, I.V., and Strømseng, A.E. (2017). Distribution and mobility of Lead (Pb), Copper (Cu), Zinc (Zn), and Antimony (Sb) from ammunition residues on shooting ranges for small arms located on mires. *Environmental Science and Pollution Research International*, 24, 10182–10196.

Martin, W.A., Lee, L.S., and Schwab, P. (2013). Antimony migration trends from a small arms firing range compared to lead, copper, and zinc. *Science of the Total Environment*, 463–464, 222–228. https://doi.org/10.1016/j.scitotenv.2013.05.086.

Mellor, A. and McCartney, C. (1994). The effects of lead shot deposition on soils and crops at a clay pigeon shooting site in northern England. *Soil Use and Management*, 10, 124–129.

Migliorini, M., Pigino, G., Bianchi, N., Bernini, F., and Leonzio, C. (2004). The effects of heavy metal contamination on the soil arthropod community of a shooting range. *Environmental Pollution*, 129, 331–340. https://doi.org/10.1016/j.envpol.2003.09.025.

Ministry of Environmental Protection of the People's Republic of China. (2018). Soil Environmental Quality, Risk Control Standard for Soil Contamination of Agricultural Land Standard (GB15618-2018). China.

Mollah, A.S. and Ferdous, M.J. (2025). Distribution of radionuclides in soil and their entry into food through uptake by plants. In: *Mechanism of Radionuclide Uptake into Food and Consequences for Humans*, Chapter 1, pp. 1–54. World Scientific Publishing Company. https://doi.org/10.1142/9789811297519_0001.

Mollah, A.S., Begum, A., Ullah, S.M., and Khan, Z.H. (2021). Studies on radionuclide transfer from soil to the food chain of man in tropical environment of Bangladesh. IAEA-TECDOC-1979 (Supplementary Files).

NIOSH (National Institute for Occupational Safety and Health). (2009). Preventing occupational exposures to lead and noise at indoor firing ranges. Department of Health and Human Service, Centers for Disease Control and Prevention. Available at: https://www.cdc.gov/niosh/docs/2009-136/pdfs/2009-136.pdf. Accessed 14 February 2025.

Park, J. and Bae, B. (2014). Uptake and transformation of RDX by perennial plants in Poaceae family (amur silver grass and reed canary grass) under hydroponic culture conditions. *Journal of Korean Society of Environmental Engineers*, 36, 237–245.

Robinson, B.H., Bischofberger, S., Stoll, A., Schroer, D., Furrer, G., Roulier, S., *et al.* (2008). Plant uptake of trace elements on a Swiss military shooting range: Uptake pathways and land management implications. *Environmental Pollution*, 153, 668–676. https://doi.org/10.1016/j.envpol.2007.08.034.

Rodríguez-Seijo, A., Alfaya, M.C., Andrade, M.L., and Vega, F.A. (2016). Copper, chromium, nickel, lead and zinc levels and pollution degree in firing range soils. *Land Degradation and Development*, 27, 1721–1730. https://doi.org/10.1002/ldr.2497.

Rodríguez-Seijo, A., Cachada, A., Gavina, A., Duarte, A., Vega, F., Andrade, M., and Pereira, R. (2017). Lead and PAHs contamination of an old shooting range: A case study with a holistic approach. *Science of the Total Environment*, 575, 367–377. https://doi.org/10.1016/j.scitotenv.2016.10.018.

Rodriguez-Seijo, A., Lago-Vila, M., Andrade, M.L., *et al.* (2016). Pb pollution in soils from a trap shooting range and the phytoremediation ability of Agrostis capillaris L. *Environmental Science and Pollution Research*, 23, 1312–1323.

Rodríguez-Seijo, A., Lago-Vila, M., Arenas-Lago, D., Andrade, M.L., and Vega, F.A. (2016). Pollution and risk assessment of potential hazardous elements in a shooting range soils (NW Spain). *Spanish Journal of Soil Science*, 6(2).

Rooney, C.P., McLaren, R.G., and Cresswell, R.J. (1999). Distribution and phytoavailability of lead in a soil contaminated with lead shot. *Water, Air, and Soil Pollution*, 116, 535–548. https://doi.org/10.1023/A:1005181303843.

Sanderson, P., Fangjie, Q.F., Seshadri, B., *et al.* (2018). Contamination, fate and management of metals in shooting range soils — a review. *Current Pollution Reports*, 4, 75–187.

Sanderson, P., Naidu, R., and Bolan, N. (2014). Ecotoxicity of chemically stabilised metal(loid)s in shooting range soils. *Ecotoxicology and Environmental Safety*, 100, 201–208.

Sanderson, P., Naidu, R., Bolan, N., Bowman, M., and Mclure, S. (2012). Effect of soil type on distribution and bioaccessibility of metal contaminants in shooting range soils. *Science of the Total Environment*, 438, 452–462. https://doi.org/10.1016/j.scitotenv.2012.08.014.

Scoriza, R.N. and Correia, M.E.F. (2019). Establishment of leguminous trees in the soil of a shooting range. *Floresta e Ambiente*, 26, e20170805.

Scott, R.I. (2001). Lead contamination in soil at outdoor firing ranges. Available at: http://www.princeton.edu/~rmizzo/firingrange.htm. Accessed 14 February 2025.

Sehube, N., Kelebemang, R., Totolo, O., Laetsang, M., Kamwi, O., and Dinake, P. (2017). Lead pollution of shooting range soils. *South African Journal of Chemistry*, 70, 21–28. https://doi.org/10.17159/0379-4350/2017/v70a4.
Shoukat, S., Nazneen, S., Khan, S., and Zafar, U. (2020). Contamination and potential ecological risk factors of potentially toxic elements present in the soil of shooting range: Comparison with the global soils. *Environment and Pollution*, 9(1).
Smith, C. (2019). Areas of major concentration in the use and traffic of small arms. In: *Small Arms Control*. Routledge, London, UK, pp. 83–126.
Sorvari, J.H. (2002). Shooting ranges as a risk to human health and environment – three case studies. SETAC Europe 12th Annual Meeting: Challenges in Environmental Risk Assessment and Modelling: Linking Basic and Applied Research. 12–16 May 2002, Vienna, Austria.
Suter II, G.W., Efroymson, R.A., Sample, B.E., *et al.* (2000). *Ecological Risk Assessment for Contaminated Sites*. Lewis Publishers, Boca Raton, p. 438.
Tariq, S.R. and Ashraf, A. (2016). Comparative evaluation of phytoremediation of metal contaminated soil of firing range by four different plant species. *Arabian Journal of Chemistry*, 9, 806–814.
Uddin, M.J., Parveen, Z., and Hossain, M.F. (2016). Status of heavy metals in water and sediments of canals and rivers around the Dhaka city of Bangladesh and their subsequent transfer to crops. *Advances in Plants & Agriculture Research*, 5(4), 593–601. https://doi.org/10.15406/apar.2016.05.00189.
United States Environmental Protection Agency (USEPA). (2005). Use of field-scale phytotechnology for chlorinated solvents, metals, explosives, and propellants, and pesticides. Phytotechnology mechanisms solid waste and emergency response (5102G), EPA 542-R-05-002.
Vantelon, D., Lanzirotti, A., Scheinost, A.C., and Kretzschmar, R. (2005). Spatial distribution and speciation of lead around corroding bullets in a shooting range soil studied by micro-X-ray fluorescence and absorption spectroscopy. *Environmental Science and Technology*, 39, 4808–4815. https://doi.org/10.1021/es0482740.
WHO (World Health Organization). (1996). Permissible limits of heavy metals in soil and plants. World Health Organization, Geneva, Switzerland.
Zhu, Y., Che, R., Tu, B., Miao, J., Lu, X., Li, J., Zhu, Y., and Wang, F. (2024). Contamination and remediation of contaminated firing ranges – an overview. *Frontiers in Environmental Science*, 12, 1352603. https://doi.org/10.3389/fenvs.2024.1352603.

https://doi.org/10.1142/9789819822249_0008

Chapter 8

Environmental Impacts of Military Base Development

Jozef Sabol

Police Academy of the Czech Republic in Prague, Prague, Czech Republic

sabol@polac.cz

Abstract

Militarism has a clear and significant effect on the natural environment. Many studies explore how military activities—especially military bases—affect ecosystems. This chapter delves into the various risks and environmental consequences linked to the operations of military bases, which may also pose threats to human health. It highlights recent research that underscores the deep connections between humans, other living beings, and the environment. By focusing on this interconnected perspective, we gain new insights into environmental issues, the long-term impacts of militarism, and potential solutions to the climate crisis. Military bases often operate under a veil of secrecy, shielded from public view. These installations serve as key military personnel, equipment, and strategic infrastructure centres. While their existence is generally known, their internal operations and environmental implications are rarely disclosed. Moreover, there is limited publicly available data on how these bases may contribute to environmental degradation and the effects on people, wildlife, and plant life. In general, the regulatory control of the situation at military bases concerning compliance with the relevant national safety and security requirements is not always adequate, which may result in excessive pollution and other hazards at the bases.

Keywords: Military bases, main powers behind, environmental impact, troops, pollution, locations, present situation, future development.

1. Introduction

Military bases considerably impact the environment, affecting the surrounding ecosystems. Their operations generate pollution, resulting in deforestation, habitat destruction, and disruptions to wildlife populations. The noise from military activities can interfere with animal behavior and migration patterns, further impacting biodiversity and local ecological balance. Waste management poses challenges, particularly with the disposal of hazardous materials, contributing to environmental degradation.

Furthermore, military bases are significant energy consumers, producing substantial greenhouse gas emissions and leaving a considerable carbon footprint. Many bases are adopting renewable energy solutions and enhancing energy efficiency to mitigate these effects. Addressing the environmental consequences of military installations is essential for minimizing harm and promoting sustainability within defence operations.

The presence of military bases has a multifaceted impact on local communities, offering both advantages and challenges. They not only contribute to economic growth and infrastructure development but also raise environmental, social, and security issues. Strategic planning, active community involvement, and sustainable policies are crucial to optimizing benefits and minimizing drawbacks. The future of military base development will be shaped by technological advancements, shifting geopolitical dynamics, sustainability considerations, and the growing demand for rapid deployment capabilities.

Military bases fulfil various functions, including serving as training facilities for soldiers and operational centres for military missions. They provide a secure environment where troops can reside and perform their duties, ensuring their readiness for combat or humanitarian operations. Moreover, these bases function as logistical centres, efficiently deploying personnel and supplies across different areas. Recognizing the wide-ranging roles of military bases is essential to understanding their importance within the overall defence structure.

Military bases are carefully structured to maximize both efficiency and security. They are divided into distinct zones based on their specific purpose. For instance, distinct areas are designated for barracks or living quarters, administrative buildings, training fields, storage facilities, and recreational spaces for personnel. The arrangement mitigates potential risks and enhances operational functionality by positioning key resources within well-protected boundaries.

Due to their sensitive nature, military bases require high security. Multiple security measures should be implemented to protect personnel and assets from potential threats. These may include perimeter fencing with controlled access points, surveillance systems such as CCTV (Closed-Circuit Tele Vision) cameras and motion detectors, armed guards at entry checkpoints, and advanced technologies, such as biometric identification. The extent of security depends on factors such as the base's location and its strategic significance.

Military bases require robust infrastructure to function effectively. This includes utilities such as electricity, water supply, sewage systems, communication networks, and transportation infrastructure. For instance, runways or helipads are essential for air operations, while roads and rail networks facilitate the movement of troops and supplies. The scale and complexity of these infrastructural elements depend on the size and purpose of the base.

The presence of military bases can have a significant impact on the environment. Extensive construction activities, training operations, and handling of hazardous substances may pose risks to nearby ecosystems. Nevertheless, numerous military bases adopt strategies to minimize these impacts.

As a whole, militaries are among the world's biggest fuel consumers, accounting for 5.5% of global emissions (Wikipedia, 2025a). They represent at least three-quarters of the global military presence, and the US armed forces are the biggest emitters. Military activity affects the physical environment in the following direct ways: pollution of the air, land, and water in peacetime, as well as the immediate and long-term effects of armed conflict. The progression of warfare from chemical weapons to nuclear weapons has increasingly created stress on ecosystems and the environment. When wars create energy shortages and rise in fossil fuel prices, it disrupts global climate action.

Human security has traditionally been solely linked to military activities and defence. Scholars and institutions are now increasingly calling for a more holistic approach to security, particularly emphasizing the interconnections and interdependencies between humans and the environment (Jorgenson *et al.*, 2016). Military activity has significant impacts on the environment. War can be destructive to the socioenvironment, and military activities produce extensive amounts of greenhouse gases (that contribute to anthropogenic climate change), pollution, and resource depletion, among other environmental impacts.

2. Locations and Potential Impact of Some Military Bases of Three Superpowers

Several major global powers maintain military presence and foothold in different regions of the world through their overseas bases, which help provide them a logistical edge over their foes. These military bases are often equipped with modern weapons and equipment, allowing countries to strike enemy targets anywhere in the world (Jawad, 2023).

The United States, for instance, has more than 750 military bases across 80 countries and has deployed nearly 175,000 soldiers in 159 countries of the world, which provides it with a key edge over foes and the capability to remain within striking distance if a conflict were to break out. These military bases are a key factor that contributes to America's global hegemony and enable Washington to counter threats emanating from countries such as China, Russia, Iran, and North Korea.

2.1. *United States*

While the United States has a substantial number of military bases on its home soil, it also maintains many bases abroad that help ensure the nation's defense and combat readiness. In particular, the vast array of allied nations in Europe (Fig. 1) and their locations across the Atlantic help maintain global military dominance (US MBE, 2024).

The US military maintains hundreds of military installations, both inside the United States and overseas. Some American bases are also NATO-led, with forces from multiple countries (Instagram, 2014; Wikipedia, 2025b). An overview of US military bases in the world is shown in Fig. 2 (Reddit, 2020).

The US controls about 750 bases in at least 80 countries worldwide and spends more on its military than the next 10 countries combined. The

Fig. 1. Main US Military bases in Europe (based on UD MBE, 2024).

US military presence worldwide can be illustrated by information given in Fig. 3 (Al Jazeera, 2020).

Fort Bragg is the largest US Army base at home by population, serving a population of 47,499 active-duty soldiers, 51,564 reserve components and temporary duty students, 16,276 civilian employees and contractors, and 71,960 active-duty family members (Fort Bragg, 2025). Fort Bragg is located just west of Fayetteville, North Carolina, known throughout its history for its cultural diversity and military presence. On the other hand, Camp Humphreys in South Korea is the largest US military base overseas camp, housing 500 buildings and amenities and having a population of 40,000—bigger than some US cities (Wikipedia, 2025c).

2.2. *Russian Federation*

During the Soviet Union time, the country had altogether 22 military bases in the following countries in Europe and Asia (based on Wikipedia, 2025d):

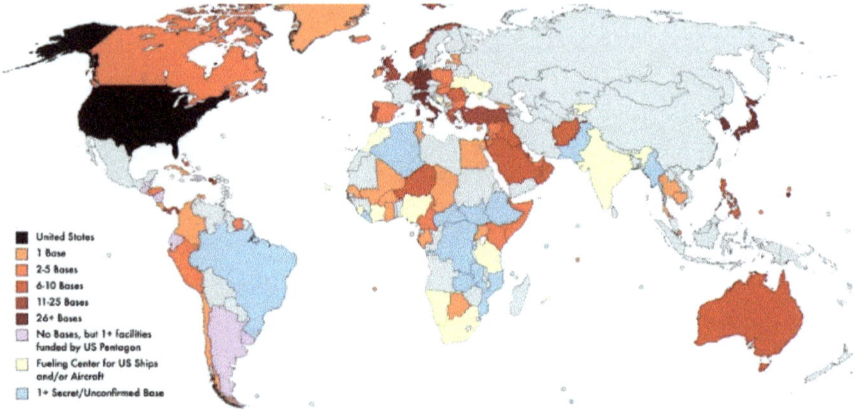

Fig. 2. A map of every United States military base in the world in 2020.

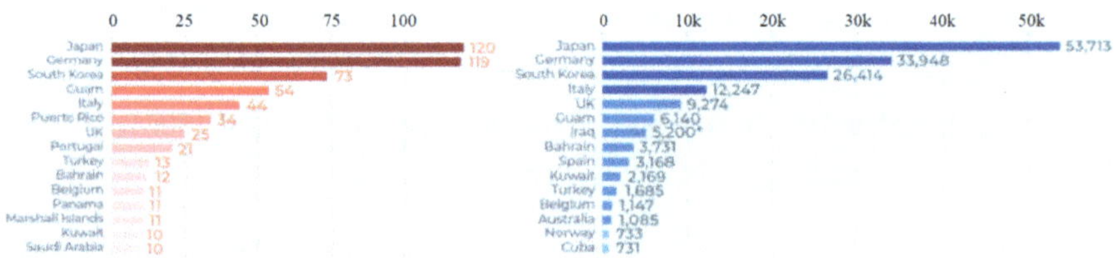

(a) (b)

Fig. 3. US military presence in terms of (a) countries with the most number of US bases and (b) countries with the most number of US troops.

Afghanistan, Albania, Austria (from 1945 to 1955), Azerbaijan (up to 2013), Baltic States (Estonia, Latvia, and Lithuania) (from 1991 to 1994), Kampuchea (from 1980 to 1992), China (from 1949 to 1953), Cuba (closed in 2002), Czechoslovakia (from 1968 to 1991), Ethiopia (between 1977 and 1991), Georgia (up to 2007), East Germany (from 1945 to 1994), Finland (from 1940 to 1941 and from 1944 to 1956), North Korea (1945–1948), Hungary (from 1944 to 1991), Kazakhstan (up 2020), Mongolia (1921–1927, 1939–1951, and 1962/1992), Poland (from 1945 to 1993), Romani (from 1944 to 1958), and Ukraine (more bases before 2014, now only in Crimea and the occupied east territory of the country).

After the breakup of the former Soviet Union, the number of military bases of the Russian Federation in other countries was reduced to 13 (Fig. 4).

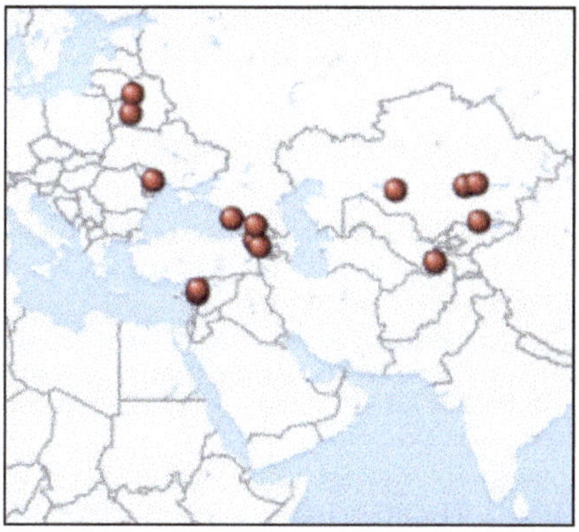

Fig. 4. Current foreign Russian military installations (excluding Crimea).

The majority of *Russia's* military bases and facilities are located in former Soviet republics, which in Russian political parlance is termed the "near abroad". Following the dissolution of the Soviet Union, many of the early-warning radar stations ended up in former Soviet republics. As of 2020, only the radar in Belarus is still rented by Russia. The list of military bases outside Russia and in some occupied territories is shown in Table 1 (based on Wikipedia, 2025).

2.3. *China*

Modern China is widely recognized as a rising global superpower, with its expanding influence playing an increasingly significant role on the world stage. The 2013 initiative to strengthen the Chinese nation highlights ambitions of asserting China's economic, political, and military dominance. A key element of this strategy involves bolstering military strength—particularly naval capabilities—and undertaking comprehensive armed forces reforms. Historically, China has prioritized the development of a strong land-based military, often placing less emphasis on naval power. However, expanding China's maritime geopolitical and economic presence has now become a central objective. In recent years, Chinese leadership has frequently employed assertive language to defend national interests and has described the early decades of the 21st century as a time of strategic opportunity (Singleton, 2024).

Amid the most extensive military expansion since the Second World War, the People's Liberation Army (PLA) is actively working to broaden

Table 1. Russian military bases outside Russia and in some occupied territories.

Country	Some details	Est. No. of personnel
Armenia	Russian Military Base in Gyumri and the Russian Airbase in Erebuni Airport near Yeraevan	3,214 to 5,000
Belarus	Russian military presence in Belarus: The Baranavichy Radar Station, the Vilyeyka Naval Communication Centre, Joint Force and Air Defense training Centre in Baranavichi	1,500
Georgia Abkhazia (disputed)	Following the Russo-Georgian War in 2008, Russia has maintained a large presence in the partially recognized states of Abkhazia and South Ossetia. The Russian Military Base is located in Abkhazia and hosts approximately 4,500 personnel.	4,500
Georgia South Ossetia (disputed)	Following the Russo-Georgian War in 2008, Russia has maintained a large presence in the partially recognized states of Abkhazia and South Ossetia. The Russian Military Base is located in South Ossetia and hosts approximately 3,500 personnel.	3,500
Kazakhstan	The Baikonur Cosmodrome is rented to Russia but is now under civilian administration. The Sary Shagan anti-ballistic missile testing range and the Kambala air base are also operated by Russia.	
Kyrgyzstan	The Air Base (military unit), the 954th test base of anti-submarine weapons (military unit), the naval communication center (military unit), and the radio-seismic laboratorv of the seismic service.	
Moldova Transnistria (disputed)	Russia maintains an operational group of forces in the Transnistria separatist region of Moldova to guard an ammunition depot at Cobasna.	1,500
Syria	Tartus naval base, Khmeimim Air Base Following the fall of the Assad regime, the future of Russian presence in Syria is uncertain.	7,000
Tajikistan	Russian Military Base, Okno space surveillance station.	7,500
Ukraine Russia (disputed)	Sevastopol Naval Base of the Black Sea Fleet, in Crimea, rented by Russia prior to the annexation of Crimea by the Russian Federation in 2014. In July 2015, Crimea had been fully integrated into Russia so the base in Sevastopol is no longer classed by Russia as overseas.	26.000

its overseas military footprint. This effort includes the operation of current PLA bases, constructing new military facilities, and identifying potential sites for future outposts.

As part of a concerted counter-basing campaign, the United States and allied governments should pro-actively engage with countries that appear most at risk of either hosting a permanent Chinese military presence or supporting the PLA's global logistics architecture to undermine China's basing pursuits.

China's strategy hinges, in part, on establishing an international network of "strategic strong points" that can provide support for overseas military operations or act as a forward base for deploying military forces overseas. The PLA's expanding global footprint and ability to conduct a wider range of missions, including limited warfighting, carry major risks for the United States and its allies in the Indo-Pacific as well as other operational theaters. Historically, the text of China's "active defense" strategy emphasized the PLA's role in defending the country's territorial integrity and winning localized wars in China's "near seas." Today, China's growing emphasis on "far seas protection" necessitates that the PLA "adapt", itself to tasks in different regions, develop the capacity of its combat forces for different purposes, and construct a combat force structure for joint operations.

The PLA's expanded mission prioritizes securing China's major trade, energy, and resource routes along its principal sea line of communication (SLOC), which runs from mainland China through the Malacca Strait and into the Indian Ocean and Gulf of Aden. China is also leveraging its civilian logistics systems and commercial infrastructure, including projects financed by Chinese companies and/or affiliated with China's Belt and Road Initiative, to support the PLA's growing access needs.

Currently, the PLA operates one declared overseas military base in Djibouti, established in 2017. Relatedly, the PLA maintains several man-made artificial islands in disputed parts of the South China Sea that it has fortified with missiles, runways, and advanced weapons systems. Beyond these naval outposts, the PLA's Strategic Support Force (SSF) also operates tracking, telemetry, and command (TT&C) stations in Pakistan, Namibia, Kenya, and Argentina that support China's space and satellite operations.

After years of denials by Cambodian and Chinese officials, the PLA appears on track to *inaugurate* its second overseas military base—and its first in the Indo-Pacific—at Cambodia's Ream Naval Base as early as 2023. The US Defense Department contends the PLA may have also made basing overtures to three additional countries in recent years: Vanuatu, the Solomon Islands, and Namibia. The Defense Department has further identified at least 13 additional locations that the PLA has likely considered to support its overseas military logistics and basing infrastructure. The Biden administration dispatched delegations to two of them—the United Arab Emirates in 2021 and Equatorial Guinea in 2022—in an attempt to dissuade both governments from hosting an official Chinese military presence.

So far, China has a limited number of its military bases in other countries but is seeking new overseas military facilities around the world to boost geopolitical aspirations (Fig. 5). The People's Liberation Army (PLA) will continue to pursue the establishment of overseas military installations and access agreements in an attempt to project power and

Fig. 5. Some of the current Chinese military bases (black) and the ambition plan in accordance with which China wants to become a strong military power.

protect China's interests abroad. Beyond developing its military base in Djibouti and its military facility at Ream Naval Base in *Cambodia*, Beijing reportedly is considering pursuing military facilities in multiple locations, including—but not limited to—Burma, Cuba, Equatorial Guinea, Pakistan, Seychelles, Sri Lanka, Tajikistan, Tanzania, and the UAE.

3. Specific Environmental Impact of Military Bases

In general, as with any human activity, activities carried out at military bases also result in some ecological consequences, which are primarily damaging due to their harmful effects on living organisms and the environment (Fig. 6).

The ecological impact is the effect that something has on living beings, i.e., organisms and their non-living environment. The term may refer to either the impact of human activities or natural events.

It is obvious that the construction and all activities at the military bases will affect the surrounding environment and influence people living close to such installations. It is evident that environmentalists have raised these problems all the time. Recently, some actions have been initiated to minimize the impact of the military bases on the pollution of air, water, soil, plants, and animals in areas close to these facilities.

Given the recent developments in diversified approaches to the study of the impacts of military bases all over the world, this field has great potential to develop via active interaction between various disciplines,

Fig. 6. Ecological impact that human activities have on living organisms and the environment (based on MBN, 2024).

looking at specific geographical locations. A historical examination is crucial in this regard, and an emphasis on generational and geographical differences provides us with a clearer understanding of the situation. Furthermore, there could be a possibility of analyzing in detail how environmental impact is understood according to the socially ordering categories of race, gender, and class in the host community.

Depending on the context, the term "military base" may refer to any establishment (usually permanent) that houses a nation's *armed forces* or even organized *paramilitary* forces such as the *police*, *constabulary*, *militia*, or *national guards*. Alternatively, the term may refer solely to an establishment used only by an *army* to the exclusion of a base used by either an *air force* or a *navy*. This is consistent with the different meanings of the word *military*.

Throughout history, military bases have had a profound impact on international relations. From the time of ancient empires to modern nation-states, their establishment and development have been shaped by factors such as geopolitical interests, security needs, and economic influences. Gaining insight into the historical context of military bases and their relationships with host nations is crucial for comprehending the complexities and challenges of base presence in the contemporary world.

Military bases are often shrouded in secrecy, hidden from the public's prying eyes. These sprawling complexes serve as the nerve centres for military operations, housing troops, equipment, and infrastructure necessary for defence and strategic planning. While their existence is widely known, these bases' inner workings and intricate details remain largely unknown to the general public. This section will further delve into the hidden world of military bases, shedding light on their purpose, organization, and the infrastructure that supports them.

Military bases serve a multitude of purposes, ranging from training grounds for soldiers to command centres for military operations. They provide a secure environment for troops to live and work, ensuring their readiness for combat or humanitarian missions. Additionally, military bases act as logistical hubs, facilitating the movement of personnel and supplies across different regions. Military bases are meticulously organized to optimize efficiency and security. They are divided into various sections or zones based on their function. For example, separate areas may be dedicated to barracks or housing facilities, administrative offices, training grounds, storage depots, and even recreational spaces for soldiers. The layout of military bases is designed to minimize vulnerabilities and enhance operational effectiveness by strategically placing critical assets within secure perimeters.

The idea of military bases can be traced back to ancient times, notably to the Roman Empire, which set up garrisons to maintain control over the lands it had conquered. These military outposts were vital for expanding the empire's reach and protecting its strategic interests. One prominent example is Hadrian's Wall in Britain, constructed by the Romans as a defensive structure and a symbol of their military power.

Throughout the colonial era, European powers set up military bases in their overseas territories to protect trade routes, exploit local resources, and exert authority over native populations. These bases often served dual roles as centers for military operations and as administrative or political headquarters. A prominent instance is the British Raj in India, which developed a network of strategically located military cantonments across the region.

In the Cold War period, military bases expanded significantly as the global superpowers aimed to extend their influence and strengthen deterrence. The United States and the Soviet Union engaged in a strategic contest to establish bases across the globe, leading to a dense network of military outposts. Key examples include the U.S. Naval base at Guantanamo Bay in Cuba and Soviet naval facilities in Vietnam.

With the conclusion of the Cold War, global power structures underwent significant shifts, prompting a reassessment of military base strategies. Many nations found it increasingly difficult to justify the ongoing presence of foreign troops without the immediate threat of communism.

As a result, host countries and stationed forces engaged in negotiations over the size, function, and longevity of military installations. A key example of this realignment is the downsizing of US military bases in Germany following its reunification.

In recent years, establishing foreign military bases has sparked debate in various regions worldwide. Host countries struggle to reconcile their security requirements, sovereignty issues, and public opinion with these bases' advantages. Military bases are critical assets that require robust defence systems to safeguard against a wide range of threats and intrusions. These installations serve as the nerve centres for military operations, housing personnel, equipment, and sensitive information. As such, they become prime targets for various adversaries seeking to disrupt military capabilities or gain access to classified resources. Understanding the types of threats and intrusions faced by military bases is crucial in developing effective defence strategies.

All military bases serve a multitude of purposes, ranging from training grounds for soldiers to command centres for military operations. They provide a secure environment for troops to live and work, ensuring their readiness for combat or humanitarian missions. Additionally, military bases act as logistical hubs, facilitating the movement of personnel and supplies across different regions. Understanding military bases' diverse roles is crucial in comprehending their significance within the broader defence framework.

Some radical steps have been taken, especially by the US Army, which in 2022 released its first Climate Strategy Implementation Plan to respond to climate change. Among other tactics, the army proposed to achieve net-zero army greenhouse gas emissions by 2050 and to provide 100% carbon-pollution-free electricity for army installations by 2030. The army should optimize fuel, water, electricity, and other resources to increase its resilience and reduce its impact on the planet (Oshiro, 2023).

Military pollution is a significant contributor to climate change. Their negligence has come at a huge environmental cost, and further reforms need to be considered to protect our planet. The main aim in the coming years should be to reduce the emission of any contaminants produced by military bases and to adopt stricter protection measures to minimize the consequences of military bases on the local communities and the environment. Maintaining military bases and deploying troops are generally energy-intensive activities. Weapons use—apart from causing human deaths and injuries—also causes considerable damage to the natural environment.

3.1. *Air pollution*

Scientists have long been aware that inhaling the air can negatively impact our health. According to the World Health Organization, over 90% of the global population is exposed to polluted air. Among the different forms of air pollution, particulate matter (PM) poses a particularly serious threat. These fine airborne particles—composed of both organic and inorganic substances and often referred to as aerosols—are linked to numerous premature deaths annually, primarily due to cardiovascular, respiratory, and other related health conditions.

Most of the main sources of air pollution are also applied to pollutants produced by military bases and their activities, which are illustrated in Fig. 7.

A significant issue concerning air pollution caused by military activities—especially the operation of military bases—is depicted in Fig. 8. This figure compares the US military's annual CO_2 (carbon dioxide) emissions with those of several entire countries. Such emissions play a major role in driving climate change. Historically, the US military has been one of the world's largest contributors to climate pollution, consuming vast amounts of liquid fuels and releasing more CO_2 than most nations.

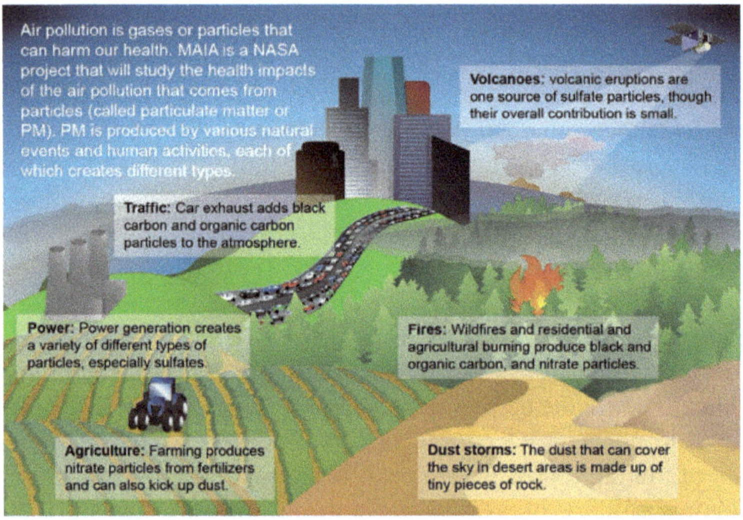

Fig. 7. Air pollution has numerous sources, both natural and human-produced (Buis, 2021).

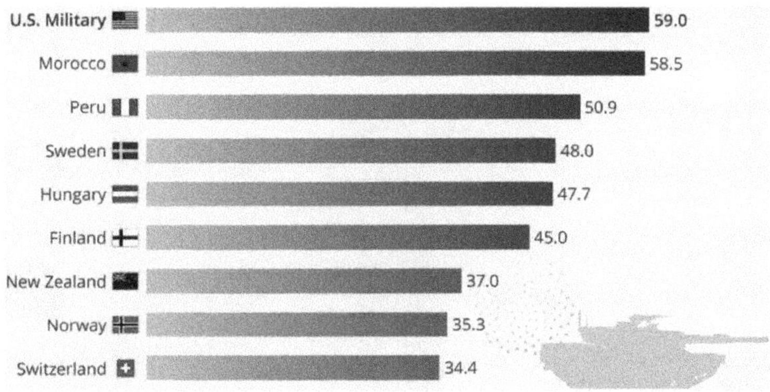

Fig. 8. Emission of CO_2 by the US military activities in 2017 (million tonnes) compared to the total emission of this pollutant by some other countries (Lewis, 2021).

Air pollution continues to be the most significant environmental threat to human health. The World Health Organization (WHO) reports that 99% of the global population lives in regions where air quality falls short of recommended levels. Globally, it ranks as the second leading cause of death and is also the second biggest risk factor for mortality in children under five—after malnutrition—because of its serious effects on respiratory and developmental health.

The relationship between air pollution and our health has been studied for decades. However, 2022 is the first year Global Action Plan's review of the evidence base, which is approved by a number of expert bodies, has been able to confirm that air pollution can have health impacts on every major organ in the body, can shorten our lives, contribute toward chronic illness, and put us more at risk from COVID-19 (Fig. 9).

Airborne particulate matter pollutants consist of a complex blend of extremely fine solid particles and liquid droplets suspended in the air. When inhaled, these particles can penetrate the lungs, leading to various health issues. Of particular concern are two categories based on particle size: (a) those with a diameter of 10 microns or less (PM10), and (b) finer particles with a diameter under 2.5 microns (PM2.5). PM2.5 has attracted growing attention because medical studies suggest these tiny particles are largely responsible for the negative health effects linked to particulate matter. Several factors influence the type and likelihood of health consequences, including particle size and chemical composition, concentration in the air, length of exposure, and individual human factors such as age, existing

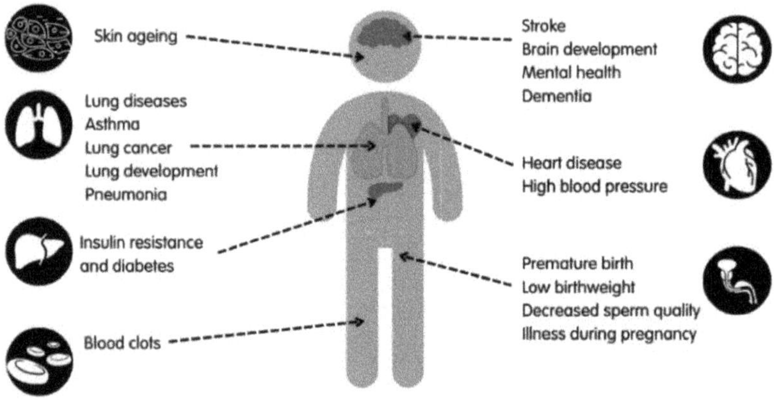

Fig. 9. Air pollution impacts most of the organs in the human body (GAP, 2022).

health conditions, overall health, and genetics. These factors, combined with gaps in scientific research, make it difficult for the medical field to accurately assess health risks for generally healthy populations, particularly since most research has focused on older or more vulnerable individuals.

3.2. *Water pollution*

Water pollution on military bases arises from a range of factors. A primary contributor is using and storing hazardous substances such as fuel and chemicals, which are frequently handled on-site. Military training activities can also introduce pollutants—such as explosives and other harmful materials—into nearby water sources. In addition, aging infrastructure often leads to leaks and spills that release toxic substances into the environment.

The impact of such pollution can be severe and long-lasting, posing serious health risks to both military personnel and nearby communities while also disrupting local ecosystems. To mitigate these dangers, military installations need to emphasize environmental stewardship and adopt more sustainable practices. The contamination of water at the military base stems from multiple sources, as shown in Fig. 10.

These hazardous compounds are often referred to as "forever chemicals" due to their long-lasting presence in the environment. The problem of polluted water at military sites is not merely a historical issue—it remains a current and pressing challenge.

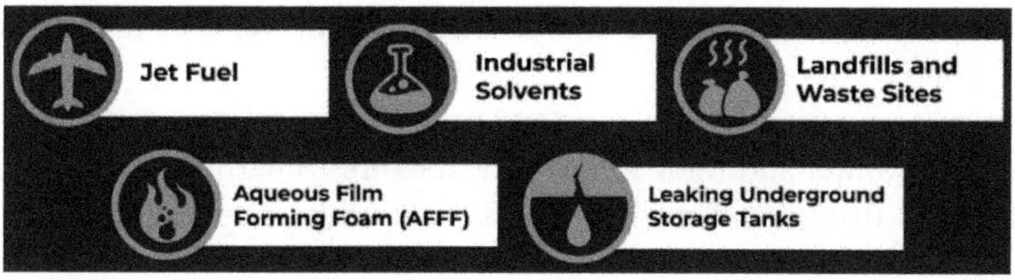

Fig. 10. Principal sources of military base contamination (King Law, 2025).

More broadly, water pollution is a pressing global concern that affects oceans, rivers, lakes, and other vital ecosystems. A significant cause of this pollution is runoff from industrial and agricultural operations, which carries various harmful chemicals and pesticides with it. These substances not only threaten aquatic organisms but can also endanger human health.

One major contributing factor to water contamination is the improper disposal of household waste and sewage, which can result in the accumulation of harmful pathogens in the water. Understanding these common causes is vital for effectively protecting water sources and preserving the species that rely on them.

Contaminants in drinking water supply pose serious risks to human health and the environment. These hazardous substances can originate from various sources, including industrial discharges, agricultural runoff, and untreated sewage. In significant quantities, these pollutants can lead to severe illnesses or even fatalities among those who consume the affected water.

Moreover, pollution can devastate aquatic ecosystems, undermining the planet's ecological balance. Ongoing research into the effects of these contaminants is essential for developing strategies to reduce their impact and safeguard both public health and the environment.

The consequences of water pollution extend beyond human health—they also threaten ecosystems. Therefore, military installations must take comprehensive measures to prevent and manage such incidents. A key prevention strategy involves the safe storage, handling, and disposal of hazardous materials in strict compliance with guidelines and regulations issued by relevant authorities.

Routine inspections of wastewater treatment facilities, fuel storage units, and other potential sources of water pollution are critical for early identification and swift resolution of any emerging problems. These preventive actions help military installations minimize their environmental footprint while safeguarding the health of both service members and nearby communities.

Water contamination on military bases poses serious health risks to military personnel and their families. Exposure to harmful chemicals and pollutants in drinking water can result in a wide range of health issues, from skin irritations to life-threatening illnesses such as cancer. Unfortunately, numerous military bases across the nation have reported problems with contaminated water supplies.

Those living on these bases must be informed about the potential dangers and take steps to protect themselves, such as using bottled water or installing home filtration systems. Protecting the health and well-being of those who serve our nation must always be a priority, and access to clean, safe drinking water is a fundamental part of that commitment.

Water contamination has far-reaching consequences beyond harming the environment—it can pose serious human and animal health risks. In these situations, seeking legal support to pursue rightful compensation and ensure those at fault are held accountable is essential. Consulting with experienced environmental law attorneys who understand the complexities of these cases can make a significant difference. Although navigating the legal system might seem daunting, taking that first step can pave the way for meaningful change and a safer, healthier future for those affected by pollution.

Testing has revealed that drinking water near military bases throughout the country contains various harmful contaminants. These pollutants, detected in groundwater at military sites nationwide, have been linked to serious health issues such as a heightened risk of cancer, reproductive problems, and potentially fatal illnesses. The primary toxic substances polluting the water at these bases are shown in Fig. 11 (King Law, 2025).

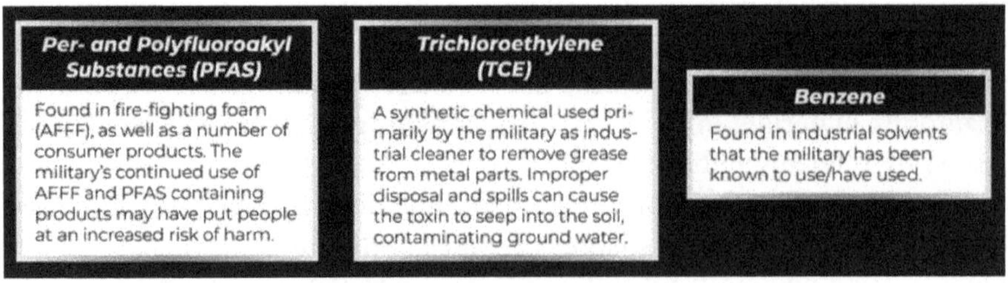

Fig. 11. Types of usual contaminants in military drinking water.

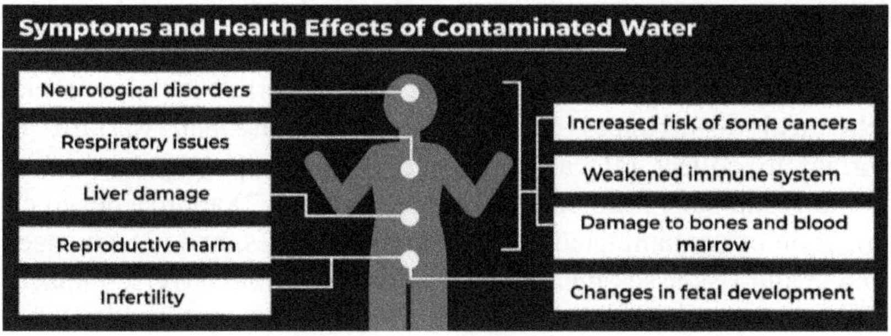

Fig. 12. Main symptoms and relevant health effects of contaminated water.

Testing of drinking water near military bases across the US has revealed the presence of multiple harmful contaminants known to negatively impact human health. While many of these so-called "forever chemicals" are found in a range of consumer and industrial products—making their exact source difficult to pinpoint—some are consistently detected at military sites.

These hazardous substances in groundwater around military facilities have been linked to serious health problems, such as cancer, reproductive issues, and other life-threatening conditions. The signs and symptoms of drinking contaminated water are often relatively mild at first. They may include frequent headaches, frequent bouts of dizziness, and feelings of nausea. Over time, with increased exposure, the symptoms will get worse, potentially becoming life-threatening. It is essential to obtain an accurate diagnosis from a medical professional. Some of these symptoms are presented in Fig. 12 (King Law, 2025).

3.3. *Land pollution*

Although crucial for national security and defence, military operations inevitably have considerable environmental effects that warrant careful scrutiny. The relationship between military activities and ecological well-being brings forward urgent concerns about sustainability and the long-term health of ecosystems.

Military base operations can lead to significant land and soil contamination due to the use of toxic chemicals, improper waste disposal, and accidents. Contaminants such as per- and polyfluoroalkyl substances (PFAS) from firefighting foams and hazardous materials from maintenance activities can leach into the environment, posing risks to both ecosystems and human health.

Military activities drastically affect soil properties mainly via physical/chemical disturbances during military training and warfare. It is important to collect relevant information about (1) physical/chemical disturbances in soils following military activities, (2) approaches to characterization of contaminated military-impacted sites, and (3) advances in human health risk assessment for evaluating potential adverse impacts.

It has been established that soils within the territory of military facilities remain significantly contaminated with toxic compounds from ammunition and their residues for many decades. Assessing the environmental consequences of military actions requires examining multiple aspects, including habitat degradation and chemical contamination. Promoting an informed discussion on this topic is essential to raising awareness and supporting approaches that integrate defence efforts with responsible environmental management. Contaminated soil can also lead to other health problems, such as depression of the central nervous system, headaches, nausea, fatigue, eye irritation, and skin rashes.

Healthy soil is foundational to human health. Healthy soil is needed to grow crops, provide food, and sustain populations. It supports diverse ecosystems and critical ecological services such as pollination. It stores water and prevents floods. It captures carbon and slows global climate change. Soil pollution is a significant and growing threat to human health. Heavy metals, organic chemicals such as pesticides, biological pathogens, and micro/nanoplastic particles may pollute soil.

One of the most dramatic ways humans can affect soil properties is through the performance of military activities, which cause disturbances to the soil. There are basically three types of soil: physical, chemical, and biological. Physical disturbances to soil include sealing due to the building of defensive infrastructures, excavation of tunnels, compaction by the traffic of machinery and troops, or cratering by bombs. Chemical disturbances consist of the input of pollutants, such as oil, heavy metals, nitroaromatic explosives, organophosphorus nerve agents, dioxins from herbicides, or radioactive elements. Biological disturbances occur as unintentional consequences of the impact on soil's physical and chemical properties or the introduction of *micro-organisms* such as botulin or anthrax. Soil represents a secure niche where such pathogens can perpetuate their virulence for decades.

Radionuclide pollution also occurs from inadequate storage or dismantling of nuclear weapons. Cases of radioactive waste releases during the Cold War were detected in and near military facilities in both the former Soviet Union (Skipperud *et al.*, 2005) and the USA (Skryness *et al.*, 1994).

Major concern is arising from the use in conflicts of depleted uranium (DU) armour-piercing weapons and their long-term effect on the environment and human health. The risk of soil pollution arises from the fine radioactive dust released during the impact of DU projectiles into armour. DU is a dense byproduct of the uranium enrichment process that creates nuclear weapons and fuel for nuclear power reactors. It has a toxicity similar to lead, and it is an alpha particle emitter, with a half-life of 4.5 billion years. DU has higher radioactivity than natural uranium.

Heavy metals, especially lead (Pb), are the major cause of pollution in the military shooting range soils. Bullets, which are primarily made of Pb, are a substantial source of this pollution. During speciation, this Pb is distributed into different metal forms. Different physicochemical properties of the soil, such as pH, moisture content, cation exchange capacity (CEC), and organic matter, play a crucial role in the Pb's distribution, transformation, and bioavailability. The concentration of Pb found in different shooting ranges is examined. Moreover, bullet weathering and the availability of contaminants in the soil are influenced by the physicochemical properties of the soil. For the management of firing range pollution, a variety of strategies have been investigated, including soil washing, phytoremediation, and chemical stabilization. This review focuses on the pollution status of different shooting ranges, the impact of the physicochemical properties of soil on the distribution, speciation, and transformation of Pb, and different mitigation strategies to control Pb pollution in military shooting ranges (Alasmary, 2025).

Military traffic, including wheeled or tracked heavy vehicles maneuvers, is another significant factor affecting the soil. Compaction is the main negative impact of military traffic on soil, significantly modifying soil's hydraulic properties and making soils more vulnerable to erosion and runoff. In saturated soils, excess loading could cause liquefaction, resulting in various issues, including mud formation (Broomandi *et al.*, 2020).

Finally, the instigation of fires in crops or forests has numerous adverse consequences on the physical properties of soil. The most indirect significant impact is susceptibility to erosion, which rapidly propagates on

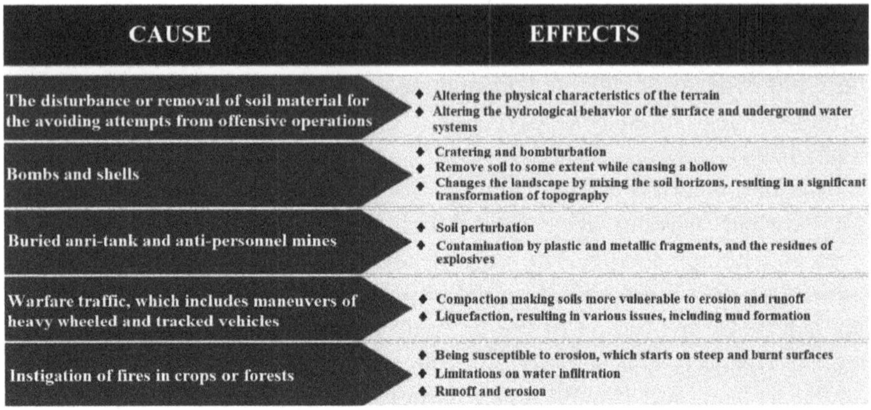

Fig. 13. Cause-and-effect diagram for physical impacts of warfare on soil (based on Broomandi *et al.*, 2020).

steep, incinerated surfaces. Furthermore, due to the fire-induced formation of a hydrophobic layer at shallow depths that prevents/limits water infiltration, these areas are highly vulnerable to runoff and erosion, which may result in shallow landslides and debris flow. A cause-and-effect diagram regarding the physical impacts of military activities on the soil is presented in Fig. 13.

The present situation regarding the environmental effect of military activities (training and warfare) on soils requires contaminated site characterization and the evaluation of the progress and challenges that impact human health risk.

Soil physical disturbances via excavation of tunnels or trenches, compaction by traffic of troops and machinery, and cratering by bombs (bombturbation) sometimes drastically affect soil, such as its topography and hydraulic properties, which may lead to landslides and erosion. Chemical disturbances via the introduction of a wide list of potentially toxic elements (PTEs), energetic compounds (ECs), and chemical warfare agents (CWAs) may adversely impact ecosystems as well as human health. Attempts to control these substances are similar to the attempts to control industrial contamination, e.g., via defining and enforcing maximum allowable environmental concentrations.

Only a few countries have established general limits and legislation addressing environmental contaminants. Even where such regulations exist, they often lack consistency, may exclude certain key pollutants, and are not always grounded in clear scientific evidence. This highlights the need for a unified, scientifically robust framework that comprehensively addresses a broad range of contaminants.

Military-affected areas frequently contain numerous potentially toxic elements (PTEs), including lead (Pb), mercury (Hg), arsenic (As), cadmium (Cd), copper (Cu), and nickel (Ni). Many of these elements—particularly lead, which is both highly toxic and persistent—are commonly detected at elevated levels. However, in assessing site contamination, the focus should be on the mobility of PTEs rather than their total concentration. While some elements, such as lead, are typically less mobile under natural conditions, shifts in environmental factors—such as changes in soil pH from acid rain or variations in redox conditions due to flooding—can significantly enhance their mobility.

Although emerging contaminants (ECs) and chemical warfare agents (CWAs) are less frequently found in soils compared to PTEs, they may still pose serious risks. A thorough understanding of their physicochemical properties—such as solubility, Henry's constant, and octanol-water partition coefficient—as well as their biodegradability is essential. These characteristics greatly influence their behavior in the environment, including how they move through and persist in soil compartments. Effective site assessments should incorporate detailed spatial and vertical profiling based on this information. While most studies include such profiling, many still overlook critical aspects such as contaminant speciation, fractionation, and mobility.

The human health risk assessment of military-impacted sites mainly followed a well-agreed framework for evaluating chemical risks. However, this framework is multi-layered, and the extent (depth) it was used varied from study to study. Contaminant chemical speciation and bioaccessibility, which directly affect the results for risk characterization, were largely not considered and thus should be adequately integrated into the assessment process for more accurate results in the future.

Military planners rarely prioritize environmental considerations when planning or executing combat operations. In times of active conflict, the focus is typically on achieving strategic objectives and maintaining superiority on the battlefield, often at the expense of environmental concerns. In contrast, developed nations tend to integrate environmental protection into military training and base operations during peacetime, largely due to legal obligations. Unfortunately, most conflicts take place in developing countries, where fragile institutions and scarce resources make it difficult to develop and enforce environmental regulations—especially after hostilities have ended. Overall, modern warfare has a predominantly negative impact on the environment and biodiversity. Although precise data is lacking, it is likely that armed conflict plays a significant role in driving environmental degradation and biodiversity loss in some regions.

Studies and evaluations have indicated that military technology advancements have indirectly positively affected ecosystem functioning in some situations. In certain cases, warfare has acted as a beneficial factor for ecosystems, as the unintentional exclusion of humans from conflict zones has created safe havens for wildlife and, at times, suitable habitats for endangered or threatened species.

Warfare generally has a detrimental impact on ecosystems' functioning, with land pollution being a major contributor through multiple pathways. Activities tied to armed conflict—such as nuclear explosions, military exercises, and the release of harmful chemicals—result in reduced populations of local plants and animals and a decline in biodiversity within the affected regions. These negative outcomes have been recorded across a range of ecosystems and species, indicating both short-term and long-lasting effects. Figure 14 outlines the typical ecological consequences linked to different types of military actions. However, in certain cases, war can inadvertently benefit the environment, as limited

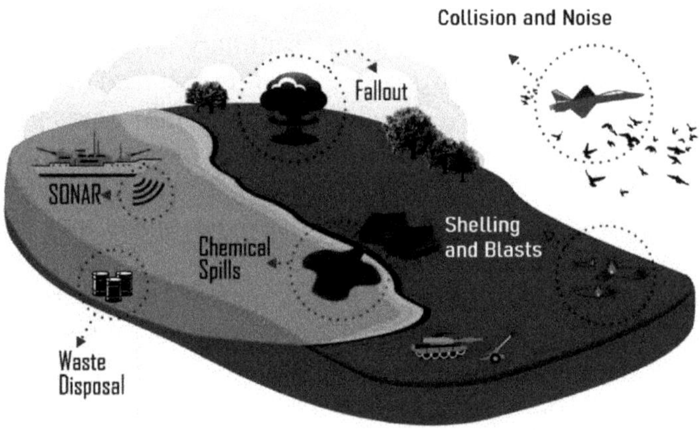

Fig. 14. Overview of the potential deleterious impacts of warfare on the environment, including terrestrial, aerial, and naval activities occurring during military base activities (based on Lawrence *et al.*, 2015).

human presence in conflict zones may provide refuges for wildlife and create habitats that support endangered or vulnerable species.

The environmental impacts of infrastructure construction vary depending on the specific location. For instance, building naval ports and shipyards typically poses a higher risk of contaminating nearby water bodies compared to constructing land-based airstrips, which are often located far from water sources and surrounded by natural vegetation that acts as a buffer. Even when constructing similar types of infrastructure, the environmental consequences can differ significantly depending on the local landscape and ecosystem, making such impacts highly dependent on the site.

3.4. *Animals and plants*

Wars and military operations, including those conducted at military bases, have significant negative effects on both people and biodiversity. Even though these consequences are not ethically justifiable, some unintended outcomes have led to positive developments in animal and plant life conservation. Examples from various regions and historical contexts demonstrate that extensive military areas and activities can be repurposed as nature reserves—provided that well-informed, science-based administrative decisions are made within a broader, coordinated strategy for biosphere conservation alongside other human activities.

Military training areas are generally large areas dedicated to various training activities and all operations necessary to prepare for war. They present an essential and viable potential for nature conservation. Indeed, they represent large continuous spaces distributed in various ecosystem types. These lands are also of ecological interest as military training can be subjected to both intensive and extensive land uses. Indeed, military grounds can be considered as undisturbed areas as they are closed land or as disturbed grounds that undergo a variety of training and, therefore, a variety of disturbances. This is the paradox of military lands or "ecological militarization". Usually, the core area is heavily impacted by military training activities, while the peripheral areas are protected from human activities such as urbanization and intensive agriculture.

Military bases turn out to be a haven for endangered species. A decision long ago by the military that working with conservationists was a better strategy than fighting them is one of the reasons why. Effects on wildlife of noise associated with military training activities, especially vehicle noise, artillery, small-arms and other blast noise, and helicopter noise, are reviewed but not effects on aquatic animals and effects of fixed-wing aircraft engine noise and sonic booms.

Vehicle noise can interfere with animal communication essential for reproduction. On the other hand, people who are afoot may exhibit stronger behavioral reactions than people in vehicles. Noise from helicopters (rotary-winged aircraft) is complex and varies, especially with the helicopter's aspects and model. Responses to helicopters from several kilometers distant are documented; however, both the reaction to helicopters and the degree to which such a response may habituate with time vary greatly. Vehicles and helicopters pose difficulty for researchers trying to separate acoustic from visual and other indications of their presence.

In fact, bases have the highest density of threatened and endangered plants and animals of any federal lands, even more than national parks, such as Yellowstone. Like many bases, they were built when the US population was a third of what it is now, and before development, they had gobbled so much forest, beach, and prairie. That makes bases a kind of biological time capsule. They're now home to more than 425 threatened and endangered species.

4. Conclusion

Although the military's primary mission is to safeguard citizens, its operations frequently result in environmental contamination that poses significant but often overlooked risks to both service members and civilians living in or around military installations. A major issue is the improper disposal of toxic materials at these sites.

While the health implications for military personnel are widely acknowledged, the dangers faced by surrounding civilian populations are sometimes downplayed. In recent decades, clear evidence has emerged showing that exposure to these hazardous substances can lead to immediate health problems, such as skin rashes, respiratory issues, and other minor ailments among service members. Of even greater concern are the long-term consequences of chronic exposure, which have been associated with serious illnesses, such as cancer, hormonal imbalances, and developmental problems.

These environmental threats stem from multiple sources. Routine maintenance of military equipment can result in harmful waste entering the environment. Weapons testing often causes chemical runoff that pollutes nearby water systems, while emissions from jets and armored vehicles contribute significantly to air pollution. Firefighting foams, extensively used in training and emergencies, further contaminate soil and groundwater. The use of burn pits to dispose of military waste has released toxic pollutants into the air, with wide-reaching environmental and health effects.

Despite being among the largest institutional consumers of fossil fuels and energy, militaries are typically exempt from international climate accords and stringent environmental regulations. Consequently, the greenhouse gas emissions generated by military activities are often omitted from national carbon accounting. Gaining a full understanding of the environmental costs tied to military pollution is essential. While the armed forces are vital to national security, reflecting on the environmental trade-offs involved in maintaining that security is important.

Military activities, including preparations for war or defence, inevitably result in considerable environmental degradation—polluting land, water, and air. The risks associated with radioactive contamination, such as those from nuclear submarines, must also be taken into account. However, as mentioned above, there is also some positive impact in preserving some endangered animals and plants.

Bibliography

Alasmary, Z. (2025). Lead (Pb) contamination in soil and plants at military shooting ranges and its mitigation strategies: A comprehensive review. *Processes*, 13(2), 345. https://doi.org/10.3390/pr13020345.

Al Jazeera. (2020). Infographic: US military presence around the world. *Al Jazeera News*. https://www.aljazeera.com/news/2021/9/10/infographic-us-military-presence-around-the-world-interactive.

Bragg, F. (2024). *Community Health Assessment*. Fort Bragg Department of Public Health, Fort Liberty, USA, NC, 28310, 1 July 2024.

Broomandi, P. *et al.* (2020). Soil contamination in areas impacted by military activities: A critical review. *Sustainability*, 12(21), 9002. https://doi.org/10.3390/su12219002.

Buis, A. (2021). Particulate matter air pollution has numerous sources, both natural and human-produced. NASA's Jet Propulsion Laboratory. https://commons.wikimedia.org/wiki/File:Particulate_matter_air_pollution_has_numerous_sources_both_natural_and_human-produced.webp. Accessed 3 April 2025.

Clouse, R. (2014). What are some examples of how large military training areas end up benefiting wildlife? Quora. https://www.quora.com/What-are-some-examples-of-how-large-military-training-areas-end-up-benefiting-wildlife. Accessed 23 March 2025.

Coleman, J. (2023). Endangered species are getting a boost from U.S. military bases. *Wild Hope*. https://www.wildhope.tv/article/endangered-species-boosted-by-military-bases/. Accessed 3 April 2025.

GAP (Global Action Plan). (2022). Clean Air Day 2022: Air pollution impacts every organ in the body. http://www.globalactionplan.org.uk/news/clean-air-day-2022air-pollution-impacts-every-organ-in-the-body. Accessed 23 March 2025.

Instagram. (2014). Military bases visualization. https://www.instagram.com/powerfulcountries/p/C22fo_kOXjc/. Accessed 4 March 2025.

IntPik. (2022). Russian military bases. https://intpik.ru/russian/military/bases/. Accessed 4 March 2025.

Jawad, A. (2023). Countries with the most overseas military bases. *Insider Monkey*. https://www.insidermonkey.com/blog/13-countries-with-the-most-overseas-military-bases-1237744/. Accessed 26 March 2025.

Jorgenson, A.K. *et al.* (2016). The temporal stability and developmental differences in the environmental impacts of militarism: The treadmill of destruction and consumption-based carbon emissions. *Sustainability Science*, 11(3), 505–514.

King Law. (2025). Military base water contamination lawsuit – March 2025 update. http://www.robertkinglawfirm.com/personal-injury/military-base-water-contamination-lawsuit/#sources-of-military-base-water-contamination. Accessed 26 March 2025.

Lawrence, M.J. *et al.* (2015). The effects of modern war and military activities on biodiversity and the environment. *Environmental Reviews*, 23(4), 443–460.

Lewis, J. (2021). US military pollution: The world's biggest climate change enabler? Brown University (USA), BS Statistical Review of World Energy. https://earth.org/us-military-pollution/. Accessed 31 March 2025.

MBN (Market Business News). (2024). What is the ecological impact? Definitions and examples. https://marketbusinessnews.com/financial-glossary/ecological-impact/. Accessed 26 March 2025.

Oshiro, A. (2023). From "footprint" to relationships: Impacts of US military base on Okinawa. *Sociology Compass*, 18(1), e13099. https://doi.org/10.1111/soc4.13099.

Price, J. (2016). Military bases serve as safe havens for endangered species. *NPR*. https://www.npr.org/2016/09/15/494127912/military-bases-serve-as-safe-haven-for-endangered-species. Accessed 31 March 2025.

Reddit. (2020). A map of every United States military base in the world in 2020. https://www.reddit.com/r/dataisbeautiful/comments/myh0lq/oc_a_map_of_every_united_states_military_base_in/?rdt=36068. Accessed 26 March 2025.

Singleton, C. (2024). Mapping the expansion of China's global military footprint. *Foundation for Defense of Democracies*. https://www.fdd.org/plaexpansion/. Accessed 23 March 2025.

Skipperud, L., Strømman, G., Yunusov, M., Stegnar, P., Uralbekov, B., Tilloboev, H., Zjazjev, G., *et al.* (2005). Plutonium contamination in soils and sediments at Mayak PA, Russia. *Health Physics*, 89(3), 255–266. https://doi.org/10.1097/01.hp.0000165449.06373.b0.

Skrynnes, R. (1994). Spoils of war. *Civil Engineering ASCE*, 64(11), 51–53.

USAPHC (US Army Public Health Center). (2019). Particulate matter (PM) air pollution exposures during military deployments. https://ph.health.mil/PHC-Resource-Library/Particulate_Matter_Factsheet_64-009-0414.pdf. Accessed 26 March 2025.

US MBE (US Military Bases Europe). (2024). US military bases in Europe. https://www.operationmilitarykids.org/us-military-bases-in-europe/. Accessed 4 March 2025.

Wikipedia. (2025a). Environmental impact of war. *Wikipedia – The Free Encyclopedia*. https://en.wikipedia.org/wiki/Environmental_impact_of_war. Last edited 26 March 2025.

Wikipedia. (2025b). List of countries with overseas military bases. *Wikipedia – The Free Encyclopedia*. https://en.wikipedia.org/wiki/List_of_countries_with_overseas_military_bases. Accessed 4 March 2025.

Wikipedia. (2025c). Camp Humphreys. *Wikipedia – The Free Encyclopedia*. https://en.wikipedia.org/wiki/Camp_Humphreys. Last edited 25 February 2025. Accessed 26 March 2025.

Wikipedia. (2025d). List of Russian military bases abroad. *Wikipedia – The Free Encyclopedia*. https://en.wikipedia.org/wiki/List_of_Russian_military_bases_abroad. Last edited 1 March 2025. Accessed 4 March 2025.

https://doi.org/10.1142/9789819822249_0009

Chapter 9

Integrated Approaches for RDX Bioremediation: Synergizing Microbial Formulations with Biostimulation Techniques

Abhishek Sharma[*,‡], **M. Aamir Khan**[†], **and Satyawati Sharma**[†]

[*]*Amity Food and Agriculture Foundation, Amity University Noida 201313, India*

[†]*Centre for Rural Development and Technology, Indian Institute of Technology Delhi, New Delhi 110016, India*

[‡]*asharma5@amity.edu*

Abstract

RDX (hexahydro-1,3,5-trinitro-1,3,5-triazine) is a widely utilized high-energy explosive known for its thermal stability, high density, and significant detonation velocity, making it a common component in various military munitions. However, its extensive use has led to environmental contamination, particularly at military installations, firing ranges, and manufacturing sites, posing a risk to ecosystems and human health. RDX is resistant to biological degradation due to its unique chemical structure, which complicates remediation efforts. Recent studies have explored integrated bioremediation strategies that synergize microbial formulations with biostimulation techniques to enhance the degradation of RDX in contaminated soils. These approaches leverage the natural metabolic pathways of micro-organisms, which can reduce nitro groups in RDX to less harmful derivatives, thereby facilitating its breakdown. The application of inert carriers, such as eggshell powder and cocopeat, has shown promise in improving microbial growth kinetics and enhancing RDX degradation rates. Furthermore, the ecological implications of RDX contamination are significant, as its presence in aquatic systems can lead to bioaccumulation and adverse effects on aquatic organisms, including reduced hatching success and larval survival. Thus, addressing RDX contamination through effective bioremediation not only mitigates environmental hazards but also protects biodiversity and public health, underscoring the urgent need for innovative remediation solutions. This chapter outlines the advancements in these integrated approaches and their potential to provide sustainable solutions for RDX contamination.

Keywords: RDX bioremediation, microbial formulations, biostimulation, environmental management.

1. Introduction

The current period of advanced industrialization and urbanization has led to the introduction of harmful and hazardous pollutants, including environmental xenobiotics, into our ecosystems (Rodríguez *et al.*, 2020). Xenobiotics are chemicals extrinsic to an organism's normal metabolism and can reach toxic concentrations in the absence of metabolism. The US Environmental Protection Agency's (USEPA) list of priority pollutants includes various xenobiotics, such as halogenated aliphatics, aromatics, nitroaromatics, chloroaromatics, pesticides, polychlorinated biphenyls, polycyclic aromatic hydrocarbons, phthalate esters, and nitrosamines (Dinka *et al.*, 2018).

The presence of xenobiotics in the environment is of great concern due to their toxic effects on surrounding biota. Both plants and animals, including humans, are exposed to xenobiotics through various routes, including contaminated water, soil, skin absorption, inhalation, and ingestion. Primary sources of xenobiotics include wastewater and residues from industries, such as paper and pulp mills, pharmaceutical and chemical plants, and textile mills. Secondary sources include drugs, chemical pesticides, fertilizers, and pharmaceutical products discharged directly or indirectly by manufacturers or consumers (Maurya, 2016).

Explosives are among the most significant soil contaminants posing grave environmental concerns. They are broadly categorized as low and high explosives based on their detonating potential. High explosives are further categorized into primary, secondary, and tertiary explosives. Secondary high explosives, such as Royal Demolition Explosive (RDX), trinitrotoluene (TNT), and High Melting Explosive (HMX), are less sensitive than primary explosives and require a booster and detonator to produce the desired high explosion (Sharma *et al.*, 2023). The extensive use of explosives in military, industrial, and civil applications has led to their penetration into surrounding soil and water bodies. They are released into the environment during manufacture, packaging, loading, use, and decommissioning of old munitions, resulting in land and aquatic environment contamination. Due to their high thermal stability, these explosives are recalcitrant and persist for longer durations, posing grave concerns for humans and the surrounding environment.

2. Explosives

Explosives are chemical compounds or mixtures that can be initiated to generate rapidly expanding gases, heat, light, and enormous pressure. The earliest explosive compound developed with charcoal, sulphur, and nitrate was possibly black powder or gunpowder. The use of nitric and sulfuric acid resulted in the production of more powerful, highly energetic nitrate esters, nitroaromatics, and nitramines classes of explosives. Depending upon their detonating potential, explosives are broadly categorized as low and high explosives (Fig. 1).

The low explosives burn slowly releasing gases in a controlled manner while high explosives are highly reactive and release a high pressure of gases on explosion. High explosives are further categorized into primary, secondary, and tertiary high explosives (Kalsi *et al.*, 2020).

1. **Primary high explosives:** Highly sensitive and easily detonated (e.g., lead azide, silver azide, and mercury fulminate).
2. **Secondary high explosives:** Less sensitive, requiring a booster and detonator (e.g., nitroglycerine, RDX, TNT, and HMX).
3. **Tertiary high explosives:** Highly stable, requiring secondary explosives as boosters (e.g., ammonium perchlorate, ammonium nitrate, and mononitrotoluene).

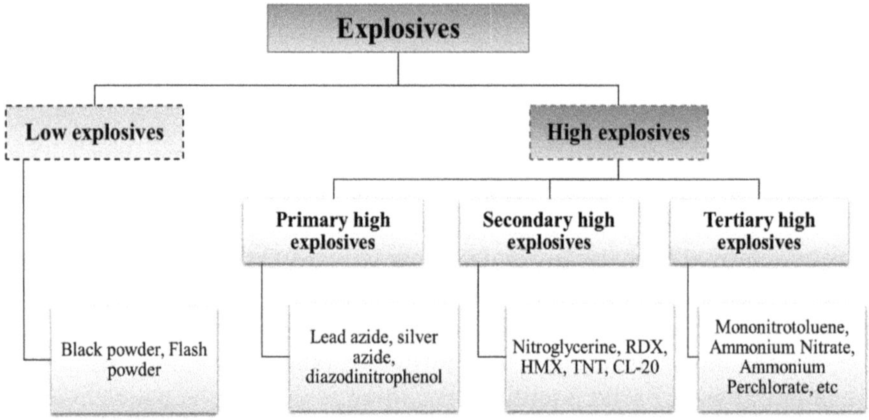

Fig. 1. A generalized scheme for the classification of explosives based on detonating potential.

The ongoing contamination of soils with explosive remnants of war not only threatens ecosystems but also poses risks to human health. As explosives degrade, they release toxic byproducts that can persist in the environment, necessitating effective monitoring and remediation strategies (Heisnam *et al.*, 2023). Recent reviews have summarized various remediation techniques, including physical, chemical, and biological methods, to address the widespread contamination caused by military operations (Mystrioti and Papassiopi, 2024). Thus, addressing explosive contamination in soils is critical for environmental protection and public health.

A pictorial view depicting the extent of explosives contamination across the world has been provided in Fig. 2. By improperly disposing of obsolete stock, manufacturing waste streams, and inadequate handling/transport, explosives are released into the environment. According to estimates, there are hundreds of explosives-contaminated sites in the United States, and even more in Europe and Asia (Broomandi *et al.*, 2020; EPA, 2021a; Pichtel, 2012). The most prevalent explosive compounds found in wastewater and soil are hexahydro-1,3,5-trinitro-1,3,5-triazine (RDX), Octahydro-1,3,5,7-tetranitro-1,3,5,7-tetrazocine (HMX), 2,4,6-trinitrotoluene (TNT), and 2,4-dinitroanisole (DNAN). RDX and TNT are classified as a class C possible human carcinogen; whereas, HMX is categorized as class D carcinogen (Mulla *et al.*, 2014; Kanekar *et al.*, 2014).

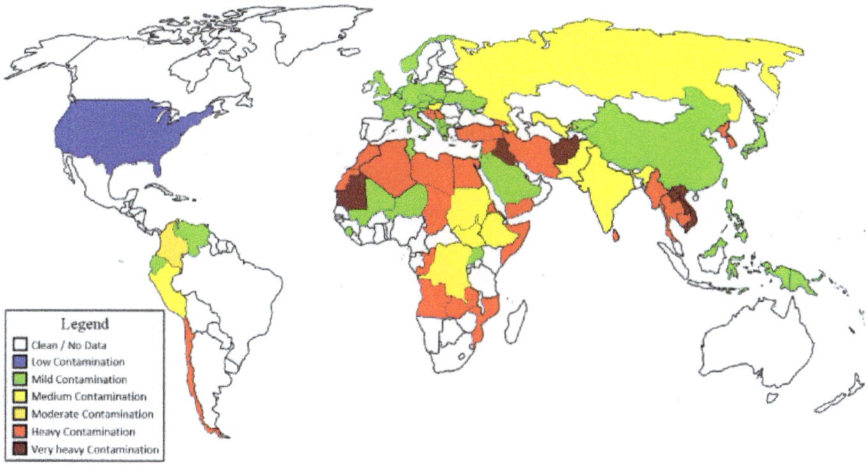

Fig. 2. Extent of explosive contamination across the world (Source: Via and Zinnert, 2016).

The explosives are chemically classified into three groups, viz. nitrate esters, nitramines, and nitroaromatics, and based on their detonating potential, explosives are categorized into low (primary) and high energy (secondary) explosives. Table 1 summarizes the different classes of widely used nitro explosive compounds with their respective structures depicting the chemical nature of each compound.

Both manufacturers and users with unregulated waste disposal and management systems are equally responsible for environmental deterioration by explosives (Ladyman *et al.*, 2019). These explosives and their derivatives have polluted the environment to the point where they endanger the health of humans, livestock, wildlife, and entire ecosystems. RDX (8800 m s^{-1}/ 338 k bar), because of its higher detonation potential than TNT (7,000 m s^{-1}/ 190 k bar), was extensively used during World War II and is reported to occur at higher concentrations in nature along with TNT (Wingfors *et al.*, 2006). Many explosive combinations which are currently in use, viz. Composition C4 (RDX 91%, plasticizer 9%), Composition A (RDX 91%, wax 9%), Composition B (RDX 60%, TNT 39%, wax 1%), and Cyclotol (RDX 75 %, TNT 25%), contain RDX as a major constituent (Pichtel, 2012), thus leading to its widespread use and release into the environment. The recalcitrance and toxic nature of RDX pose serious threat to several life forms and environment.

Table 1. List of most used nitro explosives (Khan *et al.*, 2021).

Explosive class	**Examples**	**Structure**
Nitrate esters	Propane-1,2,3-triyl trinitrate (Glycerol trinitrate (GTN))	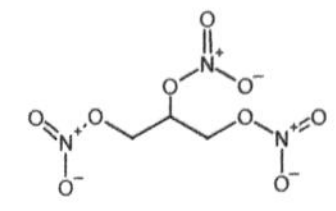
	2,2-bis(nitrooxymethyl)-propane-1,3-diyldinitrate (Pentaerythritol tetranitrate (PETN))	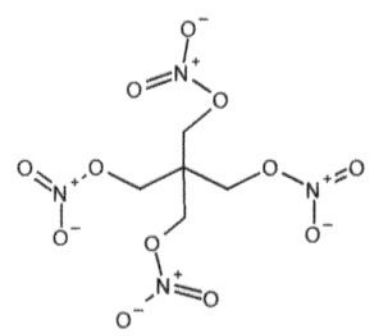
Nitramines	Hexahydro-1,3,5-trinitro-1,3,5-triazine (RDX)	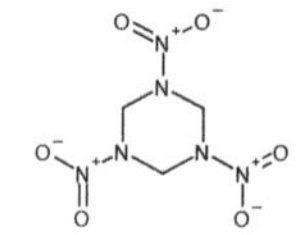
	Octahydro-1,3,5,7-tetranitro-1,3,5,7-tetrazocine (HMX)	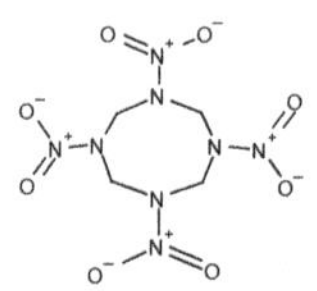
Nitroaromatics	2,4,6-trinitrophenol (TNP)	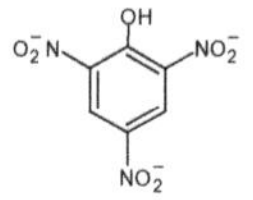
	2,4,6-trinitrotoluene (TNT)	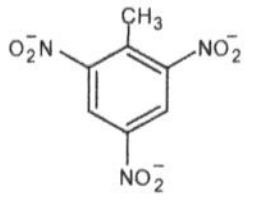
	2,4-dinitroanisole (DNAN)	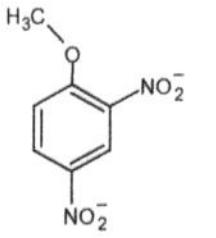

3. Royal Demolition Explosive (RDX)

The hexahydro-1,3,5-trinitro-1,3,5-triazine: Royal Demolition Explosive (RDX) is a powerful nitramine explosive (Fig. 3) and is widely used in both military and civil operations around the world. RDX, a white crystalline solid, thermally stable with high density and detonation velocity, has been extensively used in different munitions. RDX being toxic and

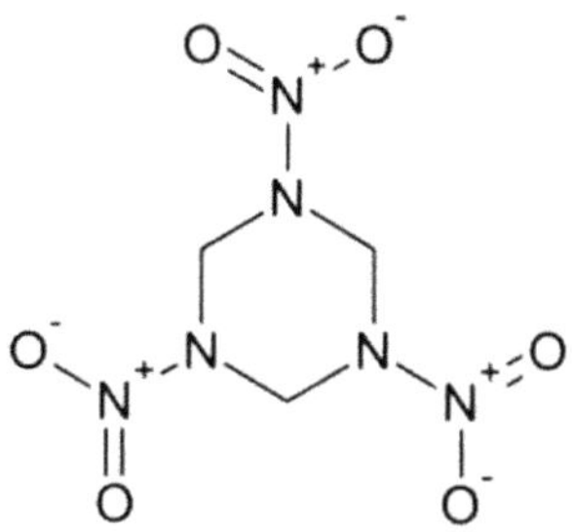

Fig. 3. Chemical structure of RDX (hexahydro-1,3,5-trinitro-1,3,5-triazine).

persistent in nature is considered an environmental hazard for soil and ground water contamination during manufacture, packaging, and decommissioning of outdated munitions (Seth-Smith *et al.*, 2002). RDX is highly toxic and listed as possible class C carcinogen by USEPA. It possesses high mobility and low degradation rates in soil due to which it quickly leaches into groundwater affecting flora and fauna surrounding the military establishments. The hazardous effects of RDX on humans includes adverse effects on central nervous system, vomiting, loss of consciousness, insomnia, anxiety, irritability, convulsions, etc. (Lee *et al.*, 2004).

The attachment of nitro groups to the central ring via N–N bonds is a unique feature of RDX structure (Fig. 3) which rarely occurs in nature and thus makes RDX resistant to biological metabolism (Khan *et al.*, 2012). RDX is gradually being considered as a toxic soil and water contaminant. Consequently, RDX and allied munitions need to be remediated from the contaminated sites; generally, the waste streams from the explosive manufacturing plant, surrounding soil, and groundwater.

The hexahydro-1,3,5-trinitro-1,3,5-triazine (RDX) was the first nitramine explosive, synthesized by Hans Hemming in 1899 (Seth-Smith, 2002). RDX is a white crystalline solid in physical appearance, which has maximum solubility of 60 mg l^{-1} at 20°C in water and is slightly soluble in solvents such as methanol and ethyl acetate. Various physical and chemical properties of RDX are listed in Table 2.

The nitramine explosive hexahydro-1, 3, 5-trinitro-1, 3, 5-triazine, or RDX (Royal Demolition Explosive), along with HMX and TNT, is one of the prominently used explosives across the world. RDX, being thermally stable with high density and detonation velocity, has been extensively used in different munitions (Lee *et al.*, 2004). RDX, comparatively mobile

Table 2. Physico-chemical properties of RDX (Abadin *et al.*, 2012).

Property	Information
Chemical identity and synonyms	Hexahydro-1,3,5-trinitro-1,3,5-triazine; cyclotrimethylenetrinitramine; Cyclonite, etc.
Chemical formula	$C_3H_6N_6O_6$
Molecular weight	222.6
Colour	White
Physical state	Crystalline solid
Melting Point	204–206°C
Density	1.82 g/ml
Solubility	Water: 38.4–38.9 mg/l (20°C); 60 mg/l (25°C); Slightly soluble in methanol, ether, ethyl acetate, glacial acetic acid
Partition coefficients	Log K_{ow} – 0.87; Log K_{oc} – 1.8

in the soil, is highly persistent with low rates of degradation and has been reported as a potential carcinogen by the US Environmental Protection Agency (USEPA), thereby causing various neurotoxic effects to aquatic as well as terrestrial organisms (Khan *et al.*, 2012).

3.1. *Environmental fate of RDX and HMX*

The high-energy explosives reach the environment during manufacturing and packing, maintenance and demilitarization, and training drills. The major explosive contaminated sites include military bases, firing ranges, burning or detonation sites, manufacturing plants, and surrounding soil (Thorn *et al.*, 2004). Several studies have been documented for the explosive's contamination across the world (Kalsi *et al.*, 2021; Best *et al.*, 2006). The maximum values detected for RDX, HMX, and TNT are as high as 74,000 mg kg^{-1}, 57,00 mg kg^{-1}, and 87,000 mg kg^{-1} soil, respectively (Best *et al.*, 2006). These explosives are chemically stable and degrade slowly in soils, so they persist for a longer period of time, inflicting harm to the environment over time (Table 3).

Table 3. Persistence of nitramine explosive RDX and HMX in different soil types under natural attenuation processes.

Explosive	Soil type	Half-life (days)	Location	References
RDX	Waimea soil: fine sandy loams	87	Waikoloa Maneuver Area, Hawaii	Alavi *et al.* (2011)
	Pantex soil: silty clay loam	60	Pantex Plant, Texas	Speitel *et al.* (2001)
	Sandy loam	188	IIT Delhi, India	Khan *et al.* (2021)
	FG sandy loam	94	Fort Greely, Washington	Jenkins *et al.* (2003)
	CG silty clay	98	Yakima Training Centre, Washington	
	YTC silty clay	154	Camp Guernsey, Wyoming	
HMX	Waimea: fine sandy loams	105	Waikoloa Maneuver Area, Hawaii	Alavi *et al.* (2011)
	Waikoloa: fine sandy loams	40		
	FG sandy loam	133	Fort Greely, Washington	Jenkins *et al.* (2003)
	CG silty clay	433	Camp Guernsey, Wyoming	
	YTC silty clay	2310	Yakima Training Centre, Washington	

Their environmental fate is determined via different abiotic and biotic factors, such as solubility and dissolution, transformation kinetics, adsorption coefficients, biodegradation, photolysis, and hydrolysis (Fig. 4). Factors such as solubility, dissolution, and adsorption determine their transport, while hydrolysis, photolysis, biodegradation, and reduction

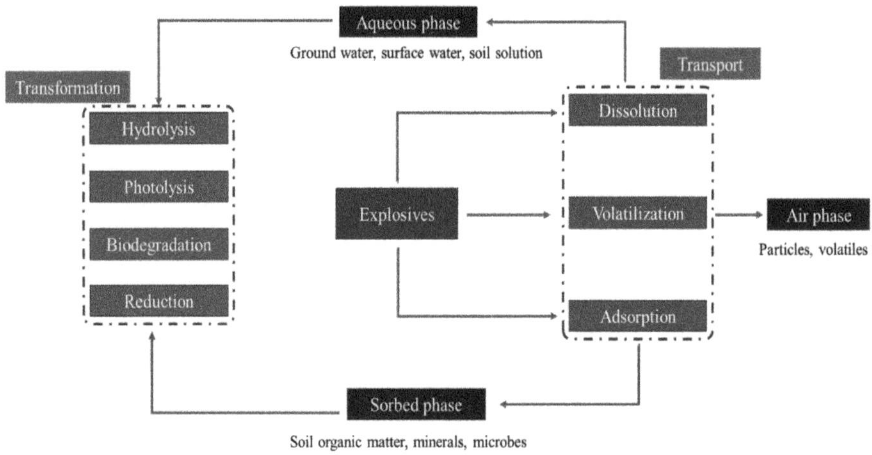

Fig. 4. Factors affecting the environmental fate of RDX (Modified from Kalderis *et al.*, 2011).

contribute to their transformation in natural environments (Kalderis *et al.*, 2011). RDX exists in the particulate phase in the atmosphere. The low solubility of RDX in water (60 mg l^{-1} at 25°C) and Henry's law constant value of 2×10^{-11}atm m^3 mol^{-1} indicate slow volatilization of RDX from water and soils (EPA, 2005). The soil adsorption coefficient values of RDX indicate moderate to high mobility of RDX in soils. TNT, on the other hand, has a higher solubility of 130 mg l^{-1} (20°C) than RDX (60 mg l^{-1}) and HMX (5 mg l^{-1}), leading to its enhanced transport toward groundwater. The soil physicochemical properties, presence of functional groups, i.e., nitro group, amino group, and environmental conditions, affect the extent of dissolution and adsorption of these explosives. Soil with high organic content showed lower dissolution and high adsorption rates, while an increase in dissolution has been observed with an increase in temperature (Brannon *et al.*, 2005). TNT transformation products 2,4-diamino-6-nitrotoluene (2,4-DANT) and 4-amino-2,6-dinitrotoulene (4-ADNT) with extra amino groups showed higher adsorption rates than TNT (Sheremata *et al.*, 1999).

Photolysis has been identified as a major transformation process for energetic compounds in surface water and soil, which acts as a major mechanism for their natural attenuation. RDX, when present in surface water, showed faster photocatalytic degradation than in particulate form (Bordeleau *et al.*, 2013). The photolysis of energetic material depends on light intensity and wavelength, particle size, and duration of light period (Kalderis *et al.*, 2011). Hydrolysis of these explosives is primarily a pH-dependent process where higher pH levels (pH > 10) are required for the transformation of functional groups, such as nitro, amine, amide, and carboxylic. The alkaline hydrolysis of RDX has been reported with the release of transformation products 4-nitro-2,4-diazabutanal (NDAB), nitrite, methanol, formaldehyde, formic acid, ammonia, and nitrous oxide (Balakrishnan *et al.*, 2003). Reduction of high explosives via iron or clay materials is another key chemical transformation process for highly oxidized nitroaromatic explosives (Uchimiya *et al.*, 2010). A rapid RDX reduction was observed in a study conducted by Naja *et al.* (2008) using Fe (0) nanoparticles as reducing agent. Iron nanoparticles when applied at the rate of 3 g l^{-1}, completely degraded 82 mmol l^{-1} of RDX within 5 mins. These chemical reactions are regulated by pH, redox potential, and catalysts, such as palladium, iron, and clay minerals (Kalderis *et al.*, 2011).

Another route for environmental transformation of high explosives is biodegradation, in which different microbial strains degrade these compounds via a complex set of enzymatic reactions in aerobic or anaerobic mode. The native microbial strains or consortia, isolated potential degraders, and engineered microbes are key players in explosives bioremediation (Rodgers and Bunce, 2001). *Pseudomonas* sp., *Raoultella planticola, Rhizobium radiobacter, Kinneretia asaccharophila, Pelomonas aquatica, Microbacterium intermedium,* and *Janibactercremeus* are a few examples of recently isolated microbial strains for explosives' degradation (Avellaneda, 2020; Ali-Begloui, 2020; Khan *et al.*, 2020; Kalsi *et al.*, 2021; Yadav *et al.*, 2020). These explosives are highly resistant to natural processes and therefore pollute water bodies and land for extended periods. Despite their recalcitrance, their continued existence is detrimental to all aquatic and terrestrial creatures, including humans. The by-products and intermediate metabolites produced by these explosives are likewise hazardous (Lotufo *et al.*, 2010).

RDX is known to be toxic to various organisms, including microbial and algal strains, plants, earthworms, and numerous aquatic and terrestrial species (Thiboutot *et al.*, 2015). The n-nitroso metabolites of RDX also present potential risks to humans and other species in the environment (Crocker and McCormick, 2016). Studies have shown that many aquatic invertebrates and vertebrates can bioaccumulate RDX when exposed to contaminated water (Liu *et al.*, 2015). Furthermore, RDX in saturated aqueous solutions has been observed to negatively impact the hatching success and larval survival of fish species during exposure tests (Lee *et al.*, 2015). The bioconcentration factor (BCF), which indicates the concentration of a chemical in an organism compared to that in surrounding water, is a critical measure of RDX's potential to accumulate in aquatic organisms (Lance *et al.*, 2020). This accumulation can lead to significant ecological risks, as it affects higher trophic levels and overall ecosystem health. According to Rosen and Lotufo (2007), the RDX BCF for the mussel *Mytilus galloprovincialis* ranged from 0.8 to 2.1. USEPA has set the soil screening levels (SSL) for RDX in residential as well as industrial soils. The SSL value for RDX is 8.3 and 38 mg/kg for residential and industrial soils, respectively (EPA, 2021b). The average BCFs in plants were 17–37 for RDX and 1 for *Eisenia fetida* (earthworms) in an exposure-based study of RDX in aged soil on various endpoint organisms representing two trophic levels (Best *et al.*, 2006).

It is imperative to clean up polluted areas due to their devastating effects on ecosystems and diverse life forms. Explosive sites cannot be utilized for other purposes until they have been cleaned up due to their toxicity, and remediation is required to keep these chemicals from migrating into groundwater. The increased public awareness of the toxicity and risk associated with energetic chemicals is further driving the development of cost-effective remediation solutions to remove them.

4. Microbial Degradation of RDX

Due to constant exposure, microbes have developed to break down RDX via various routes mediated by distinct classes of enzymes. Several microbes, viz. *Rhodococcus* sp., *Stenotrophomonas maltophila*, *Pseudomonas* sp., and *Desulfovibrio* sp., have been identified that may degrade/transform high-energy pollutants, such as RDX, while utilizing them as a sole C and/or N source or through co-metabolic activities under aerobic or anaerobic circumstances (Rodgers and Bunce, 2001). Microbes have developed to break down RDX via a variety of routes mediated by distinct classes of enzymes because of constant exposure.

The natural attenuation mechanisms in the treatment of RDX-polluted groundwater have recently been documented in evidence-based research by Fuller *et al.* (2020) in which the phylogenetic analysis confirmed the presence of potent aerobic RDX-degrading microbial groups like *Gordonia*, *Rhodococcus*, and *Methylobacterium*. The aerobic RDX degradation product 2,4-dinitrodiazabutanol (NDAB) was majorly detected in the groundwater samples (Fuller *et al.*, 2020). Another research has shown that explosive-contaminated coastal sediments are home to psychrophilic bacteria that play a crucial role in RDX breakdown. Analyses of the samples' 16S rRNA revealed that the samples contained various bacterial species from the groups *Proteobacteria, Fusobacteria, and Clostridiales*. In addition, the NDAB and MEDINA ring-cleavage products were identified (Zhao *et al.*, 2004). Natural attenuation is the simplest and most cost-effective method, but it is a slow process that takes a long time to clean up contaminated areas. Bioaugmentation and biostimulation are essential to complement these processes.

The microorganisms with explosives degrading capability when added to the contaminated sites may speed up the RDX biodegradation process in where native bacteria are unable to break down the pollutant. Environment, soil type, pH, moisture, pollutant content, and natural micro-organisms have a significant impact on the augmented microbes (Cycon *et al.*, 2017). Moshe *et al.* (2010) provided insights into the impact of soil type on RDX degradation via the compound specific isotopic analysis of RDX. The higher RDX degradation rates were observed in soil layers with higher clay content with complete mineralization within 13 days, whereas soil layers with lesser clay content took 40–60 days to degrade RDX.

A faster biotic RDX degradation by *G. metallireducens* cells was observed at pH 6.8 and 7.9 with >90% RDX removal within 30 hours compared to pH 9.2 where only 30–40% RDX degradation was observed after 30 hours. The microbial strains are usually obtained from polluted locations and administered as single cultures or consortia. While augmenting soil remediation techniques is not new, it has been extensively described in the literature (Table 4; Binks *et al.*, 1995; Bhushan *et al.*, 2002; Zhao *et al.*, 2003; Seth-Smith *et al.*, 2002; Thompson *et al.*, 2005). Recently, several RDX degrading bacterial strains, viz. *Janibactercremeus, Microbacteriumesteraromaticum, Planomicrobiumflavidum*, have been isolated from explosives contaminated sites (Kalsi *et al.*, 2020; Yadav *et al.*, 2020; Khan *et al.*, 2020).

In a recent report by Sharma *et al.* (2021), a consortium of *Paenibacillus aestuarii* and *Arthrobacter subterraneus* has shown 80.4% RDX removal potential within 12 days of incubation. On the other hand, *M. esteraromaticum* cometabolically degraded 63.9 % RDX in comparable research by Yadav *et al.* (2020). *J. cremeus* originally isolated from an

Table 4. Micro-organisms used for the degradation of RDX.

Microbial strains	Concentration tested	% Biotransformation	Products	Reference
Rhodococccus rhodochrous strain 11Y	45 μM	100% within 21 hours	Nitrite, formaldehyde	Seth-Smith *et al.* (2002)
Clostridium sp. HAW-G3	104 μM	100% within 150 hours	MNX, DNX, TNX, methanol, formaldehyde, and N_2O	Zhao *et al.* (2003)
Stenotrophomonas maltophilia PB1	100 μM 250 μM 300 μM	100% within 150 hours 100% within 200 hours 100% within 200 hours	Methylene-N -(hydroxymethyl)-hydroxylamine-N*-(hydroxymethyl) nitroamine	Binks *et al.* (1995)
Aspergillus niger	86 μM	100% within 160 hours	MNX, MEDINA, nitrous oxide, formaldehyde, ammonium ion	Bhushan *et al.* (2002)
Rhodococcus sp. strain DN22	35 μM	100% within 20 hours	Nitrite, nitrate, MEDINA, NDAB, carbon dioxide	Annamaria *et al.* (2010)
Williamsia sp. KTR4 *Gordonia* sp. KTR9	180 μM	100% within 20 hours	Nitrite, formaldehyde, and NDAB	Thompson *et al.* (2005)
Morganella morganii B2	330 μM	100% within 30 days	Nitroso derivatives of RDX and HMX	Kitts *et al.* (1994)
Providencia rettgeri B1		100% within 30 days		
Citrobacter freundii NS2		~88% after 45 days		
Acetobacterium malicum	29 μM	100% within 14 days	Nitrous oxide	Adrian and Arnett (2004)

Table 4. (*Continued*)

Microbial strains	Concentration tested	% Biotransformation	Products	Reference
Rhodococcus sp. DN22	35 μM	100% within 14 hours	Nitrite, nitrous oxide, formaldehyde	Fournier *et al.* (2002)
Klebsiella pneumoniae SCZ1	106 μM	100% within 20 days	Methanol, formaldehyde, methanol, carbon dioxide, nitrous oxide	Zhao *et al.* (2002)
Pelomonas aquatica	135 μM	80% within 240 hours	NDAB, nitrite	Khan *et al.* (2020)
Microbacteriumeste raromaticum	135 μM	70% within 216 hours	MEDINA and N-methyl-N, N′- dinitro-methanediamine, MNX, DNX	Yadav *et al.* (2020)
Janibactercremeus	270 μM	88% within 30 days	bis-(hydroxyl methyl) nitramine (BHNA), methylene-N-(hydroxy-methyl)-hydroxylamine-N-(hydroxymethyl) nitroamine	Kalsi *et al.* (2020)
Pseudomonas fluorescens I-C *Pseudomonas fluorescens* II-B	22 μM	100% within 100 hours ~65% within 340 hours	MEDINA, NDAB	Fuller *et al.* (2009)

explosive manufacturing and testing facility was evaluated for its ability to degrade RDX in aqueous phase. The isolate used RDX as a nitrogen source and eliminated 80% of it from aqueous medium in 30 days (Kalsi *et al.*, 2020).

When applied to explosives-contaminated soil, chitin increased microbial metabolism and nitrification activity, resulting in improved RDX and 2,4-dinitrotoluene (DNT) breakdown (Crocker *et al.*, 2019). In another study, waste glycerol was introduced as a biostimulant to RDX-contaminated water in monitoring wells, causing reduced conditions and organic carbon concentrations to become accessible to local microbes. When compared to control wells, the process indicated a rapid decrease in RDX levels leading to its complete mineralization (Jugnia *et al.*, 2019).

4.1. *Pathways for microbial degradation of RDX*

Micro-organisms with the capacity to optimize the breakdown of contaminants may have a specific enzyme or enzymes that degrade them into compounds of lower molecular weight. A list of RDX-degrading enzymes was published in 2015 by Sabir, including nitrate reductase (*Aspergillus niger*), oxygen-sensitive and insensitive nitroreductase (*Enterobacter cloacae* strain 96-3), and xenobiotic reductases (*Pseudomonas putida* II-b) (Sabir, 2015). There are two main routes for RDX biotransformation: sequential reduction of nitro groups and direct ring splitting. In the first pathway, the nitro (NO_2) groups are successively reduced to generate hexahydro-1-nitroso-3,5-dinitro-1,3,5-triazine or mononitroso RDX (MNX), hexahydro-1,3-dinitroso-5-nitro-1,3,5-triazine or dinitroso RDX (DNX), and hexahydro-1,3,5-trinitroso-5-nitro-1,3,5-triazine or trinitroso RDX (TNX). MEDINA, NDAB, and Bis-(hydroxylmethyl) nitramine (BHNA) are produced via direct ring breakage of RDX (Zhao *et al.*, 2004). They are subsequently processed by enzymes into simpler compounds, such as methanol (CH_3OH), formaldehyde (HCHO), hydrazine (N_2H_4), ammonia (NH_3), nitrogen oxide (N_2O), nitrite (NO_2^-), and carbon dioxide (CO_2) (Fournier *et al.*, 2002). The major reactions involved in RDX transformation include (a) two-electron reduction, (b) single-electron reduction or denitration, and (c) direct enzymatic ring cleavage (Crocker *et al.*, 2006).

(a) **Two-electron reduction:** The reduction of nitro groups on RDX initiates with the addition of two electrons ($2e^-$) leading to the formation of nitroso derivatives of RDX (mononitroso RDX (MNX), dinitroso RDX (DNX), trinitroso RDX (TNX)), aminotriazine, and hydroxyl amino derivatives of RDX (Yadav *et al.*, 2020; Zhao *et al.*, 2003). The nitro groups ($N—NO_2$) in RDX are sequentially reduced to nitroso (N—NO) functional groups. This two-electron reduction is a major RDX transformation pathway in *Methylobacterium* spp., *Shewanella* sp. HAW-EB2 and a minor route in *Klebsiellapneumoniae* strain SCZ-1, *Clostridiumbifermentans* strain HAW-1, *Shewanella* sp. strain HAW-EB4 and HAW-EB5 (Fournier *et al.*, 2005; Van Aken *et al.*, 2004).

(b) **Dinitration or single electron reduction:** The transfer of single electron to RDX generates RDX anion radical (RDX^-) which leads to the formation of ring cleavage products MEDINA and NDAB (Fuller *et al.*, 2009; Zhao *et al.*, 2002). The formation of MEDINA and/or NDAB depends on (a) the type of C—N bond that is cleaved after denitration and (b) the RDX undergoes single denitration or double denitration before ring cleavage. The single denitration of RDX produces MEDINA and/or NDAB, while the loss of two nitrite groups before ring cleavage produces NDAB as a major intermediate metabolite (Jackson *et al.*, 2007). The removal of $—NO_2$ groups from RDX could possibly occur via two different mechanisms:

i. **H-abstraction reaction:** In this reaction, the methylene group ($—CH_2—N—NO_2$) undergoes H• atom abstraction to produce a carbonyl-centered RDX free radical which after mono-denitration and di-denitration produces pentahydro-3,5-dinitro-1,3,5-triazacyclohex-1-ene and 1-nitro-1,3,5-triaza-cyclohex-3,5-ene, respectively, as intermediates (Fig. 5). These intermediates produce unstable hydroxylamine derivatives which undergo hydrolysis at —NH—CH stretch and ultimately release nitrite, ammonia, methanol, and formic acid (Halasz and Hawari, 2011; Guengerich, 2001).
ii. **Reduction of $—N—NO_2$ group:** This reaction involves transfer of an electron to nitro group of RDX ($—N—NO_2$) followed by formation of RDX radical anion (Fig. 6). Denitration of this radical anion produces aminyl radical which further undergoes H-abstraction via denitrohydrogenation to form an unstable intermediate ultimately leading to the release of MEDINA, methanol and ammonia (Halasz and Hawari, 2011).

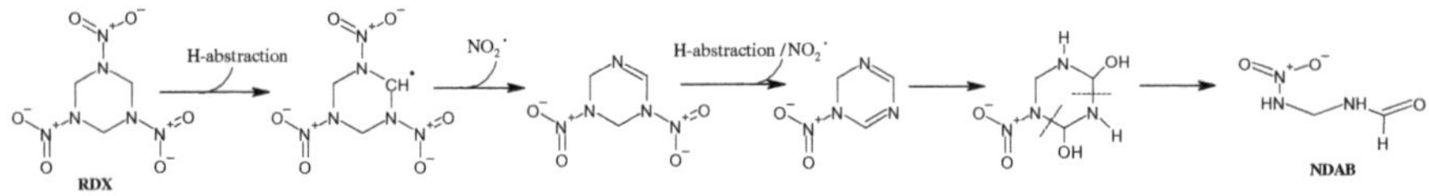

Fig. 5. Formation of NDAB via hydrogen atom abstraction.

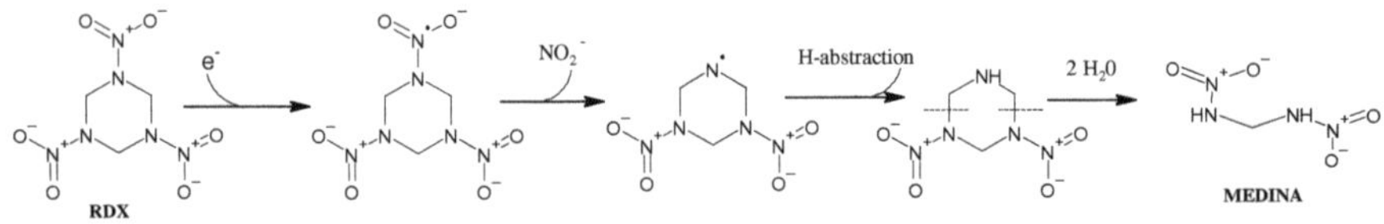

Fig. 6. Formation of MEDINA via reduction of nitro groups.

(c) Enzymatic ring cleavage: The RDX is enzymatically attacked at three different regions, viz. C—N, C—H, and N—N bonds, via different classes of enzymes, including cytochrome P450 oxygenase, hydrolase, and oxidoreductase. A list of major RDX-degrading enzymes is provided in Table 5. The site of cleavage determines the types of intermediate metabolites released during RDX degradation (Hawari *et al.*, 2000).

Several strains of *Rhodococcus* sp. (*Rhodococcus* sp. DN22, *R. rhodochrous* 11Y, and *Rhodococcus* strain 11HY) and *Gordonia* sp. KTR9 produce the cytochrome P450 oxygenase (xpl A) and oxidoreductase (xpl B) enzymes that are responsible for denitration, ring cleavage, and mineralization (Coleman *et al.*, 2002; Bhushan *et al.*, 2003). Different possible routes for RDX biodegradation are shown in Figure 7. The primary metabolites produced by strain DN22 under aerobic degradation of RDX include nitrite, nitrate, MEDINA, NDAB, and carbon dioxide (Annamaria *et al.*, 2010). Type I nitroreductase secreted by *E. cloacae* strain 96-3 and *Morganella morganii* strain B2 degrades RDX via two-electron reductive pathway (Kitts *et al.*, 2000). According to research, *P. fluorescens* strains I-C and II-B break down RDX via xenobiotic reductase enzymes Xen B and Xen A. When the strains were fed with both carbon and nitrogen sources together with RDX, Xen B degraded RDX more quickly than Xen A. In the breakdown route, MEDINA was a prominent intermediate metabolite while NDAB was another (Fuller *et al.*, 2009). Yadav *et al.* (2020) identified MEDINA and N-methyl-N, N′-dinitromethanediamine

Table 5. List of key RDX degrading genes.

Gene	Enzyme	Organism	Reference
xplA	Flavodoxin–cytochrome P450	*Rhodococcus rhodochrous* *Gordonia sp.* *Williamsia sp.*	Seth-Smith *et al.* (2002) Thompson *et al.* (2005)
diaA	Diaphorase	*Clostridium kluyveri*	Chakraborty *et al.* (2008)
xenA	Flavin mononucleotide-containing oxidoreductases	*Pseudomonas putida*	Fuller *et al.* (2009)
xenB		*Pseudomonas fluorescens*	
nfsI	Type I oxygen-insensitive nitroreductases	*Enterobacter* sp. *Morganella*sp.	Kitts *et al.* (2000)
pnrB		*Pseudomonas* sp. *Stenotrophomonas maltophilia*	Lee *et al.* (2017) Lee *et al.* (2013)

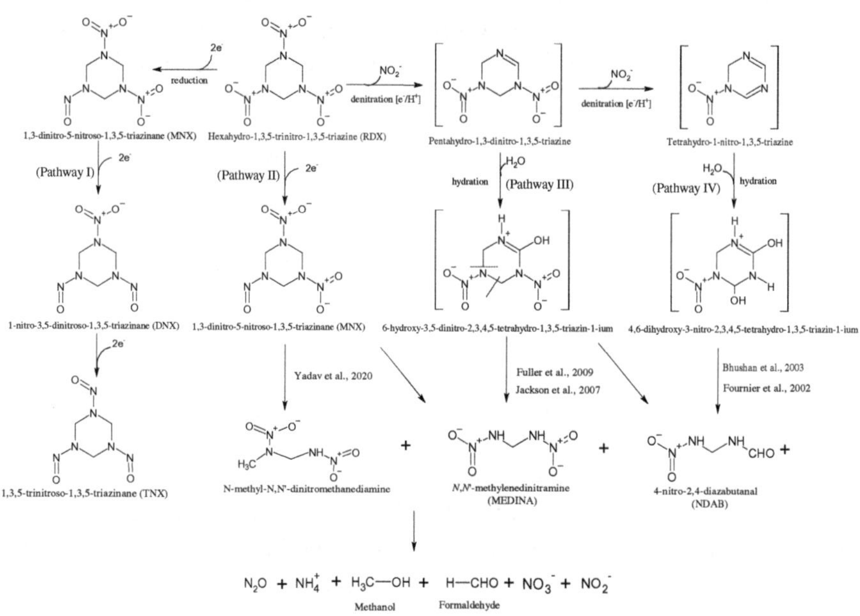

Fig. 7. Different routes for biodegradation of RDX (Yadav *et al.*, 2020; Fuller *et al.*, 2009; Jackson *et al.*, 2007; Bhushan *et al.*, 2003; Fournier *et al.*, 2002).

as RDX ring cleavage products and MNX, DNX as RDX nitroso derivates when treated with *M. esteraromaticum*. Recently, *J. cremeus* aerobically degraded RDX via single electron transfer pathway. The ring cleavage products bis-(hydroxyl methyl) nitramine (BHNA) and a derivative of MEDINA viz. methylene-N-(hydroxy-methyl)-hydroxylamine-N-(hydroxymethyl) nitrosamine was detected. Degradation led to the formation of TNX, triamino and diamino-RDX, showing that there is more than one pathway for the breakdown of RDX (Kalsi *et al.*, 2020).

5. Microbial Formulation Technology

One of the tailbacks in the successful application of the microbial remediation approach is the dearth of an efficient mechanism that could improve the growth kinetics of microbial metabolism in the remediation site (Ghosal *et al.*, 2016). We may tackle the concerns outlined above by using microbial cells in conjunction with inert carriers and other recipients, a process referred to as formulation technology. Microbes are protected from direct exposure to harsh circumstances and competing species by giving them with a microphysical habitat composed of carriers, nutrient sources, binders, and other components that enable them to reactivate and adapt to their new environment (Sharma *et al.*, 2021). Initially, the research concentrated on the mass multiplication of micro-organisms, a necessary component and a prelude to the formulation development utilizing conventional and non-traditional substrates (Preininger *et al.*, 2018; Jain and Saxena, 2019). But how to incorporate the massive amounts of cells produced into a commercially feasible formulation is the actual dilemma. As a result, we have attempted to delve into the specifics of work that has highlighted solutions relevant to our problem. In recent research, *Microbacterium esteraromaticum* cells were formulated as granules with talcum powder, alginic acid, and acacia gum. In soil microcosm experiments, the formulated *M. esteraromaticum* exhibited a 9% increase in RDX degradation over fresh culture (Yadav *et al.*, 2020). Using egg-shell powder and cocopeat as an inert carrier, Kalsi *et al.* (2020) were able to immobilise *J. cremeus* in soil that had been exposed to RDX. Soil RDX adsorption and water holding capacity have been improved because of the cocopeat. After 35 days of treatment in unsaturated and saturated soil, the formulated *J. cremeus* showed a 62% and 73% RDX degradation, respectively.

Although not much use of the immobilized microbial cells has been documented for the clean-up of carbofuran-contaminated media. Table 6 lists a few examples of the immobilization technique being used in carbofuran removal. Two specific white-rot fungi, *Phlebia* sp. and *Lenzites betulinus*, when immobilized on wheat straw, increased the degradation of carbofuran by 30% compared to the un-immobilized system (Wang *et al.*, 2020). Priyani *et al.* (2018) isolated two different microbial strains DN1

Table 6. Different immobilized microbial cultures used in bioremediation.

S. No.	Carriers for Immobilization	Contaminant	Active ingredient	References
1.	Talcum powder	RDX	*M. esteraromaticum*	Yadav *et al.* (2020)
2.	Eggshell powder, coco peat	RDX	*J. cremeus*	Kalsi *et al.* (2020)
3.	Wheat straw	Carbofuran	*Phlebia*.sp *Lenzites betulinus*	Wang *et al.* (2020)
4.	Calcium alginate beads	Carbofuran	Laccase extracted from *Lenzites betulinus*	Wang *et al.* (2017)
		Methylparathion	*Bacillus* sp.	Sreenivasulua and Aparna (2001)
5.	Polyurethane foam, agar, polyacrylamide, Alginate, and bentonite–alginate–PAC	Carbofuran	*Klebsiella pneumoniae* ATCC13883T	Kadakol *et al.* (2011)
6.	Corn cob	Carbofuran	*Burkholderia cepacia* PCL3	Plangklang and Reungsang (2013)
7.	Flyash: soil: molasses (15:3:1)	Endosulfan	Bacterial and fungal consortium	Abraham and Silambarasan (2014)
8.	Zeolite	Atrazine	*Pseudomonas* sp. strain ADP	Stelting *et al.* (2012)

(*Continued*)

Table 6. (*Continued*)

S. No.	Carriers for Immobilization	Contaminant	Active ingredient	References
9.	Plant fibres (Loofah sponge)	Methylparathion	Bacterial consortium	Moreno-Medina *et al.* (2014)
		Carbendazim		Pattanasupong *et al.* (2004)
10.	Sugarcane bagasse	Anthracene	*Phanerochaete chrysosporium*	Mohammadi and Nasernejad (2009)
11.	Tezontle	Propanil	Bacterial consortium	Herrera-González *et al.* (2013)
12.	Cloth sachet, agar, calcium alginate, and silicone tubes	Lindane	*Actinobacteria*	Saez *et al.* (2012)

and OR2 from slaughter waste house, which were further immobilized within sodium alginate and were checked for carbofuran degradation. The encapsulated microbial isolates degraded carbofuran to carbofuran phenol, which was also utilized as a source of carbon and energy, thereby decreasing the carbofuran phenol concentration by 30.37 and 32.09%, respectively. Microbial cell immobilization may provide an enhanced survival rate by stabilizing cells under a stressed environment, which enables a rapid and efficient degradation in comparison to free live cells (Moslemy *et al.*, 2002).

6. Use of Integration of Microbial Formulation and Biostimulation: A Case Study

The case study presented here is based on the work of Khan *et al.*, published in 2021, titled "Enhancing remediation of RDX-contaminated soil by introducing microbial formulation technology coupled with biostimulation." This study focused on the application of microbial formulations and biostimulants to improve the degradation of RDX in contaminated environments, particularly in soil and groundwater microcosms.

In this investigation, the researchers utilized the Gram-negative, aerobic bacterium *Kinneretia asaccharophila* strain 12853, which was cultivated in a sterile nutrient medium containing RDX at a concentration of 45 μM. The study demonstrated that this strain could effectively degrade RDX through denitrification, achieving a reduction of 75.7% in RDX levels in minimal salt medium (MSM) after 48 hours of incubation. The metabolic byproducts of this degradation process included nitrite, methylenedinitramine (MEDINA), and 4-nitro-2,4-diazabutanal (NDAB), which were confirmed through mass spectrometry analyses. The research highlighted the significant role of biostimulants in enhancing microbial activity. The addition of fructose as a biostimulant to sediment and groundwater microcosms under aerobic conditions led to improved RDX degradation rates. Furthermore, the study explored the impact of agricultural waste products, such as glycerol and wheat straw, as carbon sources to stimulate microbial growth and activity. The results indicated that when waste glycerol was applied at lower concentrations (0.5 g kg^{-1} soil), it positively influenced RDX degradation rates, achieving a mineralization rate of 55% compared to 40% in unstimulated controls.

The researchers also examined the stability and viability of the microbial strains under various environmental conditions, recognizing that factors such as temperature and moisture could significantly affect their performance in the field. To enhance the survival of the microbial formulations, the study proposed the use of stabilizing agents, such as alginate beads, which could protect the bacteria from environmental stressors and prolong their effectiveness in degrading RDX.

In addition to biological methods, the study acknowledged the limitations of conventional remediation techniques, such as soil washing and incineration, which often pose environmental risks and incur high costs. The integration of microbial bioremediation with these traditional methods was suggested as a more sustainable approach to effectively address RDX contamination. The findings from Khan *et al.* (2021) underscored the urgent need for innovative and effective remediation strategies for RDX-contaminated sites, particularly in areas surrounding military installations and ammunition plants. The successful application of microbial formulations and biostimulants demonstrated a viable pathway for mitigating the environmental challenges posed by RDX and similar explosives.

7. Conclusion and Way Forward

The bioremediation of RDX-contaminated sites presents significant environmental and health challenges due to the compound's recalcitrant nature and classification as a possible human carcinogen. This chapter outlined the potential of integrating microbial formulations with biostimulation techniques as an effective approach for RDX degradation. The microbial strains, such as *Kinneretia asaccharophila*, demonstrated significant efficacy in degrading RDX when cultured under specific conditions. The addition of biostimulants, including fructose and agricultural waste products such as glycerol, has shown to enhance microbial activity and improve degradation rates in contaminated soils and waters.

Despite the promise of biodegradation, challenges remain in optimizing these processes for field applications. Stability and viability of microbial strains under diverse environmental conditions, such as varying temperatures and moisture levels, require consideration and development of stabilization techniques, such as the use of alginate beads, to protect microbial populations from environmental stressors. Moreover, integrating these biological methods with conventional remediation techniques offers a more sustainable and effective solution to address the persistent problem of explosives contamination in soils and groundwater. This integrated approach could potentially reduce the environmental risks and costs associated with traditional remediation methods such as incineration and soil washing, providing a comprehensive strategy for managing contaminated sites.

Looking ahead, several critical areas require further exploration to enhance the effectiveness of integrated RDX bioremediation strategies: First, research into optimizing microbial formulations for greater resilience and adaptability to diverse environmental settings is crucial. This includes exploring different microbial strains and consortia that can operate synergistically under real-world conditions. Second, there is an opportunity to innovate in the use of biostimulants, particularly cost-effective and sustainable options, such as those derived from agricultural waste. Testing new biostimulant combinations could further improve RDX degradation rates and support microbial activity in the field.

Third, long-term field trials and ecological assessments are necessary to evaluate the ecological impacts and feasibility of these bioremediation strategies in various contaminated sites. Such studies would provide valuable insights into the scalability of laboratory successes.

In conclusion, advancing research in these areas holds significant promise for developing effective and environmentally conscious solutions to the enduring issue of RDX contamination. Through collaboration between scientific research, technological innovation, and environmental management, we can mitigate the risks posed by explosive pollutants and contribute to sustainable ecosystem restoration.

Bibliography

Abadin, H., Ingerman, L., and Smith, C. (2012). Toxicological profile for RDX. Centers for Disease Control and Prevention. https://stacks.cdc.gov/view/cdc/11804.

Abraham, J. and Silambarasan, S. (2014). Biomineralization and formulation of endosulfan degrading bacterial and fungal consortiums. *Pesticide Biochemistry and Physiology*, 116, 24–31.

Adrian, N.R. and Arnett, C.M. (2004). Anaerobic biodegradation of RDX by *Acetobacterium malicum* strain HAAP-1. *Current Microbiology*, 48, 332–340.

Alavi, G., Chung, M., Lichwa, J., *et al.* (2011). The fate and transport of RDX in volcanic soils of Hawaii: A laboratory and modelling study. *Journal of Hazardous Materials*, 185, 1600–1604.

Ali-Begloui, M., Salehghamari, E., Sadrai, S., Ebrahimi, M., Amoozegar, M.A., and Salehi-Najafabadi, A. (2020). Biotransformation of trinitrotoluene (TNT) by newly isolated slight halophilic bacteria. *Microbiology*, 89(5), 616–625.

Annamaria, H., Manno, D., Strand, S.E., *et al.* (2010). Biodegradation of RDX with *Rhodococcus* sp. strain DN22. *Environmental Science & Technology*, 44, 9330–9336.

Avellaneda, H., Arbeli, Z., Teran, W., and Roldan, F. (2020). Transformation of TNT, 2,4-DNT, and PETN by *Raoultellaplanticola* M30b and *Rhizobium radiobacter* M109 and exploration of the associated enzymes. *World Journal of Microbiology and Biotechnology*, 36(12), 1–14.

Balakrishnan, V.K., Halasz, A., and Hawari, J. (2003). Alkaline hydrolysis of RDX, HMX, and CL-20: Degradation pathways. *Environmental Science & Technology*, 37(9), 1838–1843.

Best, E.P.H., Geter, K.N., Tatem, H.E., and Lane, B.K. (2006). Effects, transfer, and fate of RDX from aged soil in plants and worms. *Chemosphere*, 62(4), 616–625.

Bhushan, B., Halasz, A., Spain, J., Thiboutot, S., Ampleman, G., and Hawari, J. (2002). Biotransformation of RDX catalyzed by NAD(P)H: Nitrate oxidoreductase from *Aspergillus niger*. *Environmental Science & Technology*, 36, 3104–3108.

Bhushan, B., Trott, S., Spain, J.C., Halasz, A., Paquet, L., and Hawari, J. (2003). Biotransformation of hexahydro-1,3,5-trinitro-1,3,5-triazine (RDX) by a rabbit liver cytochrome P450: Insight into the mechanism of RDX biodegradation by *Rhodococcus* sp. strain DN22. *Applied and Environmental Microbiology*, 69(3), 1347–1351.

Binks, P.R., Nicklin, S., and Bruce, N.C. (1995). Degradation of hexahydro-1,3,5-trinitro-1,3,5-triazine (RDX) by *Stenotrophomonas maltophilia* PB1. *Applied and Environmental Microbiology*, 61(4), 1318–1322.

Bordeleau, G., Martel, R., Ampleman, G., and Thiboutot, S. (2013). Photolysis of RDX in the context of military training ranges. *Chemosphere*, 93(1), 14–19.

Brannon, J.M., Price, C.B., Yost, S.L., Hayes, C., and Porter, B. (2005). Comparison of environmental fate and transport processes of explosives in saline and freshwater systems. *Marine Pollution Bulletin*, 50(3), 247–251.

Broomandi, P., Guney, M., Kim, J.R., and Karaca, F. (2020). Soil contamination in areas impacted by military activities: A critical review. *Sustainability*, 12(21), 9002.

Chakraborty, S., Sakka, M., Kimura, T., and Sakka, K. (2008). Cloning and expression of a *Clostridium kluyveri* gene responsible for diaphorase activity. *Bioscience, Biotechnology, and Biochemistry*, 72(3), 735–741.

Coleman, N.V., Spain, J.C., and Duxbury, T. (2002). Evidence that RDX biodegradation by *Rhodococcus* strain DN22 is plasmid-borne and involves a cytochrome P-450. *Journal of Applied Microbiology*, 93(3), 463–472.

Crocker, F.H. and McCormick, N.G. (2016). Complete degradation of hexahydro-1,3,5-trinitro-1,3,5-triazine (RDX) by a co-culture of *Gordonia* sp. KTR9 and *Methylobacterium* sp. JS178. *Remediation Journal*, 26(2), 5–20. https://doi.org/10.1002/rem.21457.

Crocker, F.H., Indest, K.J., and Fredrickson, H.L. (2006). Biodegradation of cyclic nitramine explosives RDX, HMX, and CL-20. *Applied Microbiology and Biotechnology*, 73(2), 274–290.

Crocker, F.H., Jung, C.M., Indest, K.J., Everman, S.J., and Carr, M.R. (2019). Effects of chitin and temperature on sub-Arctic soil microbial and fungal communities and biodegradation of hexahydro-1,3,5-trinitro-1,3,5-triazine (RDX) and 2,4-dinitrotoluene (DNT). *Biodegradation*, 30(5), 415–431.

Cycoń, M., Mrozik, A., and Piotrowska-Seget, Z. (2017). Bioaugmentation as a strategy for remediation of pesticide-polluted soil: A review. *Chemosphere*, 172, 52–71.

Dinka, D.D. (2018). Environmental xenobiotics and their adverse health impacts: A review. *Journal of Environment Pollution and Human Health*, 6(3), 77–88.

EPA (Environmental Protection Agency). (2005). EPA Handbook on the Management of Munitions Response Actions. United States Environmental Protection Agency, Washington, DC.

EPA (Environmental Protection Agency). (2021a). Fact sheet: hexahydro-1,3,5-trinitro-1,3,5-triazine (RDX). https://www.epa.gov.

EPA (Environmental Protection Agency). (2021b). Regional screening levels (RSLs) - generic tables. https://www.epa.gov.

Fournier, D., Halasz, A., Spain, J., Fiurasek, P., and Hawari, J. (2002). Determination of key metabolites during biodegradation of RDX with *Rhodococcus* sp. strain DN22. *Applied and Environmental Microbiology*, 68(1), 166–172.

Fournier, D., Trott, S., Hawari, J., and Spain, J. (2005). Metabolism of the aliphatic nitramine 4-nitro-2,4-diazabutanal by *Methylobacterium* sp. strain JS178. *Applied and Environmental Microbiology*, 71(8), 4199–4202.

Fuller, M.E., McClay, K., Hawari, J., Paquet, L., Malone, T.E., Fox, B.G., and Steffan, R.J. (2009). Transformation of RDX by xenobiotic reductases XenA and XenB. *Applied Microbiology and Biotechnology*, 84(3), 535–544.

Fuller, M.E., van Groos, P.G.K., Jarrett, M., Kucharzyk, K.H., Minard-Smith, A., Heraty, L.J., and Sturchio, N.C. (2020). Application of a multiple lines of evidence approach to document natural attenuation of hexahydro-1,3,5-trinitro-1,3,5-triazine (RDX) in groundwater. *Chemosphere*, 250, 126210.

Ghosal, D., Ghosh, S., Dutta, T.K., and Ahn, Y. (2016). Current state of knowledge in microbial degradation of polycyclic aromatic hydrocarbons (PAHs): A review. *Frontiers in Microbiology*, 7, 1369.

Guengerich, F.P. (2001). Common and uncommon cytochrome P450 reactions related to metabolism and chemical toxicity. *Chemical Research in Toxicology*, 14(6), 611–650.

Halasz, A. and Hawari, J. (2011). Degradation routes of RDX in various redox systems. In: *Aquatic Redox Chemistry*. American Chemical Society, pp. 441–462.

Hawari, J., Halasz, A., Sheremata, T., Beaudet, S., Groom, C., Paquet, L., *et al.* (2000). Characterization of metabolites during biodegradation of hexahydro-1,3,5-trinitro-1,3,5-triazine (RDX) with municipal anaerobic sludge. *Applied and Environmental Microbiology*, 66(6), 2652–2657.

Heisnam, D., Shankar, S., Chandra, D., Goel, D., Mishra, A., and Kushwaha, M. (2023). Explosive contamination in soil: Sources, environmental concerns, and phytoremediation. In: *Xenobiotics in Urban Ecosystems*. Springer, pp. 241–260.

Herrera-González, V.E., Ruiz-Ordaz, N., Galíndez-Mayer, J., Juárez-Ramírez, C., Santoyo-Tepole, F., and Montiel, E.M. (2013). Biodegradation of the herbicide propanil, and its 3,4-dichloroaniline by-product in a continuously operated biofilm reactor. *World Journal of Microbiology and Biotechnology*, 29(3), 467–474.

Jackson, R.G., Rylott, E.L., Fournier, D., Hawari, J., and Bruce, N.C. (2007). Exploring the biochemical properties of the unusual explosive-degrading P450 system XplA/B. *Proceedings of the National Academy of Sciences*, 104(43), 16822–16827.

Jain, R. and Saxena, J. (2019). Impact assessment of microbial formulations in agricultural soil. In: *Microbial Interventions in Agriculture and Environment*. Springer, Singapore, pp. 471–495.

Jenkins, T.F., Bartolini, C., and Ranney, T.A. (2003). Stability of explosives in soil. Cold Regions Research and Engineering Lab, Hanover, NH.

Jugnia, L.B., Manno, D., Dodard, S., Greer, C.W., and Hendry, M. (2019). Manipulating redox conditions for in situ bioremediation of RDX in groundwater. *Science of the Total Environment*, 676, 368–377.

Kadakol, J.C., Kamanavalli, C.M., and Shouche, Y. (2011). Biodegradation of carbofuran phenol by *Klebsiella pneumoniae* ATCC13883T. *World Journal of Microbiology and Biotechnology*, 27(1), 25–29.

Kalderis, D., Juhasz, A.L., Boopathy, R., and Comfort, S. (2011). Soils contaminated with explosives: environmental fate and evaluation of state-of-the-art remediation processes (IUPAC technical report). *Pure and Applied Chemistry*, 83(7), 1407–1484.

Kalsi, A., Celin, S.M., and Sharma, J.G. (2020). Aerobic biodegradation of high explosive hexahydro-1,3,5-trinitro-1,3,5-triazine by *Janibactercremeus* isolated from contaminated soil. *Biotechnology Letters*, 42(11), 2299–2307.

Kalsi, A., Celin, S.M., Bhanot, P., Sahai, S., and Sharma, J.G. (2021). A novel eggshell-based bioformulation for remediation of RDX contaminated soil. *Journal of Hazardous Materials*, 401, 123346.

Kanekar, P.P., Sarnaik, S.S., Dautpure, P.S., Patil, V.P., and Kanekar, S.P. (2014). Bioremediation of nitroexplosive waste waters. In: *Biological Remediation of Explosive Residues*. Springer, Cham, pp. 67–86.

Khan, M.A., Sharma, A., Yadav, S., Celin, S.M., Sharma, S., Noureldeen, A., and Darwish, H. (2021). Enhancing remediation of RDX-contaminated soil by microbial formulation technology coupled with biostimulation. *Journal of Environmental Chemical Engineering*, 9(5), 106019.

Khan, M.A., Yadav, S., Sharma, R., Dalela, M., Celin, S.M., Sharma, A., and Sharma, S. (2020). Augmentation of stimulated *Pelomonas aquatica* dispersible granules enhances remediation of RDX-contaminated soil. *Environmental Technology & Innovation*, 17, 100594.

Khan, M.I., Lee, J., and Park, J. (2012). Microbial degradation and toxicity of hexahydro-1,3,5-trinitro-1,3,5-triazine. *Journal of Microbiology and Biotechnology*, 22(10), 1311–1323.

Kitts, C.L., Cunningham, D.P., and Unkefer, P.J. (1994). Isolation of three hexahydro-1,3,5-trinitro-1,3,5-triazine-degrading species of the family Enterobacteriaceae from nitramine explosive-contaminated soil. *Applied and Environmental Microbiology*, 60(12), 4608–4611.

Kitts, C.L., Green, C.E., Otley, R.A., Alvarez, M.A., and Unkefer, P.J. (2000). Type I nitroreductases in soil enterobacteria reduce TNT

(2,4,6-trinitrotoluene) and RDX (hexahydro-1,3,5-trinitro-1,3,5-triazine). *Canadian Journal of Microbiology*, 46(3), 278–282.

Ladyman, M.K., Piperakis, M.M., Fawcett-Hirst, W., Gutierrez-Carazo, E., and Coulon, F. (2019). Decision framework for management of explosive contaminated land. *Science of the Total Environment*, 690, 730–738.

Lance, E.J. and Hurst, J. (2020). Multiple stressors in multiple species: Effects of different RDX soil concentrations and differential water-resourcing on RDX fate, plant health, and plant survival. *PLOS ONE*, 15(6), e0234166. https://doi.org/10.1371/journal.pone.0234166.

Lee, B.U., Choi, M.S., Kim, D.M., and Oh, K.H. (2017). Genome shuffling of *Stenotrophomonas maltophilia* OK-5 for improving degradation of RDX. *Current Microbiology*, 74(2), 268–276.

Lee, B.U., Choi, M.S., and Oh, K.H. (2013). Comparative analysis of explosive RDX-induced proteomes in the *Pseudomonas* sp. HK-6 wild-type strain and its rpoH mutant strain. *Biotechnology and Bioprocess Engineering*, 18(6), 1224–1229.

Lee, J., Kim, H., and Park, S. (2015). Removal of RDX using lab-scale plug flow constructed wetlands planted with Miscanthus sacchariflorus (Maxim.) Benth. *Journal of Soil and Groundwater Environment*, 20(6), 85–92. https://doi.org/10.7857/jsge.2015.20.6.085.

Lee, S.Y. and Brodman, B.W. (2004). Biodegradation of RDX. *Journal of Environmental Science and Health, Part A*, 39(1), 61–75.

Liu, Y., Zhang, Y., and Wang, J. (2015). Anaerobic biodegradation of RDX and HMX with different co-substrates. *Chinese Journal of Chemical Engineering*, 23(1), 1–7. https://doi.org/10.1016/j.cjche.2015.01.004.

Lotufo, G.R., Gibson, A.B., and Yoo, J.L. (2010). Toxicity and bioconcentration evaluation of RDX and HMX. *Ecotoxicology and Environmental Safety*, 73(7), 1653–1657.

Maurya, P.K. and Malik, D.S. (2016). Bioaccumulation of xenobiotics compound of pesticides in riverine system and its control technique: A critical review. *Journal of Industrial Pollution Control*, 32(2).

Moreno-Medina, D.A., Sanchez-Salinas, E., and Ortiz-Hernandez, M. (2014). Removal of methyl parathion and coumaphos pesticides by a bacterial consortium immobilized in *Luffa cylindrica. Revista Internacional de Contaminación Ambiental*, 30(1), 51–63.

Moshe, S.S.B., Ronen, Z., Dahan, O., Bernstein, A., Weisbrod, N., Gelman, F., and Adar, E. (2010). Isotopic evidence of in situ RDX biodegradation. *Soil Biology and Biochemistry*, 42(8), 1253–1262.

Moslemy, P., Neufeld, R.J., and Guiot, S.R. (2002). Biodegradation of gasoline by gellan gum-encapsulated bacterial cells. *Biotechnology and Bioengineering*, 80(2), 175–184.

Mulla, S.I., Talwar, M.P., and Ninnekar, H.Z. (2014). Bioremediation of 2,4,6-trinitrotoluene explosive residues. In: *Biological Remediation of Explosive Residues*. Springer, Cham, pp. 201–233.

Mystrioti, C. and Papassiopi, N. (2024). A comprehensive review of remediation strategies for soil and groundwater contaminated with explosives. *Sustainability*, 16(3), 1–14.

Naja, G., Halasz, A., Thiboutot, S., Ampleman, G., and Hawari, J. (2008). Degradation of hexahydro-1,3,5-trinitro-1,3,5-triazine (RDX) using zerovalent iron nanoparticles. *Environmental Science & Technology*, 42(12), 4364–4370.

Pattanasupong, A., Nagase, H., Sugimoto, E., Hori, Y., Hirata, K., Tani, K., *et al.* (2004). Degradation of carbendazim and 2,4-dichlorophenoxyacetic acid by immobilized consortium on loofa sponge. *Journal of Bioscience and Bioengineering*, 98(1), 28–33.

Pichtel, J. (2012). Distribution and fate of military explosives in soil: A review. *Applied and Environmental Soil Science*, 2012, 617236.

Plangklang, P. and Reungsang, A. (2013). Biodegradation of carbofuran in sequencing batch reactor augmented with immobilised *Burkholderiacepacia* PCL3 on corncob. *Chemistry and Ecology*, 29(1), 44–57.

Preininger, C., Sauer, U., Bejarano, A., and Berninger, T. (2018). Concepts and applications of foliar spray for microbial inoculants. *Applied Microbiology and Biotechnology*, 102(17), 7265–7282.

Priyani, N., Pratiwi, D., and Suryanto, D. (2018). The potency of local bacterial isolates encapsulated within sodium alginate in carbofuran degradation. In: *IOP Conference Series: Earth and Environmental Science*, 130(1), 012012. IOP Publishing.

Rodgers, J.D. and Bunce, N.J. (2001). Treatment methods for remediation of nitroaromatic explosives. *Water Research*, 35(9), 2101–2111.

Rodríguez, A., Castrejón-Godínez, M.L., Salazar-Bustamante, E., Gama-Martínez, Y., Sánchez-Salinas, E., Mussali-Galante, P., *et al.* (2020). Omics approaches to pesticide biodegradation. *Current Microbiology*, 77(4), 545–563.

Rosen, G. and Lotufo, G.R. (2007). Toxicity of explosives to *Mytilus gallopro-vincialis*. *Ecotoxicology and Environmental Safety*, 68(2), 228–236.

Sabir, D. (2015). Microbial adaptations towards utilisation of the explosive RDX in soil. Doctoral dissertation, University of York.

Saez, J.M., Benimeli, C.S., and Amoroso, M.J. (2012). Lindane removal by pure cultures of immobilized actinobacteria. *Chemosphere*, 89(8), 982–987.

Seth-Smith, H.M., Rosser, S.J., Basran, A., Travis, E.R., Dabbs, E.R., Nicklin, S., and Bruce, N.C. (2002). Cloning and characterization of RDX degradation gene cluster. *Applied and Environmental Microbiology*, 68(10), 4764–4771.

Sharma, A., Sharma, P., Celin, S.M., Rai, P.K., and Sangwan, P. (2021). Degradation of high energetic material RDX by a microbial consortium using response surface methodology. *SN Applied Sciences*, 3(1).

Sharma, K., Sharma, P., and Sangwan, P. (2023). Bioremediation of RDX and HMX contaminated soil employing a biochar-based bioformulation. *Carbon Research*, 2(1).

Sheremata, T.W., Thiboutot, S., Ampleman, G., Paquet, L., Halasz, A., and Hawari, J. (1999). Fate of TNT and its metabolites in soil. *Environmental Science & Technology*, 33(22), 4002–4008.

Speitel, G.E., Engels, T.L., and McKinney, D.C. (2001). Biodegradation of RDX in unsaturated soil. *Bioremediation Journal*, 5(1), 1–11.Sreenivasulu, C. and Aparna, Y. (2001). Bioremediation of methylparathion by free and immobilized cells of *Bacillus* sp. isolated from soil. *Bulletin of Environmental Contamination and Toxicology*, 67(1), 98–105.

Stelting, S., Burns, R.G., Sunna, A., Visnovsky, G., and Bunt, C.R. (2012). Immobilization of *Pseudomonas* sp. strain ADP for atrazine bioremediation. *Applied Clay Science*, 64, 90–93.

Thiboutot, S., Beaudin, J., and Dubeau, H. (2015). Deposition of PETN following the detonation of seismoplast plastic explosive. *Propellants, Explosives, Pyrotechnics*, 40(5), 617–623. https://doi.org/10.1002/prep.201500019.

Thompson, K.T., Crocker, F.H., and Fredrickson, H.L. (2005). Mineralization of the cyclic nitramine explosive RDX. *Applied and Environmental Microbiology*, 71(12), 8265–8272.

Thorn, K.A., Thorne, P.G., and Cox, L.G. (2004). Alkaline hydrolysis of 2,4,6-trinitrotoluene. *Environmental Science & Technology*, 38(7), 2224–2231.

Uchimiya, M., Gorb, L., Isayev, O., Qasim, M.M., and Leszczynski, J. (2010). One-electron standard reduction potentials of nitroaromatic and cyclic nitramine explosives. *Environmental Pollution*, 158(10), 3048–3053.

Van Aken, B., Yoon, J.M., and Schnoor, J.L. (2004). Biodegradation of nitro-substituted explosives by *Methylobacterium* sp. *Applied and Environmental Microbiology*, 70(1), 508–517.

Via, S.M. and Zinnert, J.C. (2016). Impacts of explosive compounds on vegetation. *Environmental Pollution*, 208, 495–505.

Wang, X., Li, Z., Yao, M., Bao, J., and Zhang, H. (2020). Degradation of carbofuran in contaminated soil by plant-microorganism combined technology. *Journal of the Serbian Chemical Society*, 85(1), 111–123.

Wang, X., Liu, L., Yao, M., Zhang, H., and Bao, J. (2017). Degradation of carbofuran by immobilized laccase. *Polish Journal of Environmental Studies*, 26(3), 1305.

Wingfors, H., Edlund, C., Hägglund, L., Waleij, A., Sjöström, J., Karlsson, R.M., Leffler, P., Qvarfort, U., Ahlberg, M., Thiboutot, S., Ampleman, G., Martel, R., Duvalois, W., Creemers, A., and Van Ham, N. (2006). Evaluation of contamination by explosives and metals in soils. FOI Report, FO1-R-1877-SE.

Yadav, S., Sharma, A., Khan, M.A., Sharma, R., Celin, M., Malik, A., and Sharma, S. (2020). Enhancing RDX remediation through water-dispersible *Microbacterium esteraromaticum* granules. *Journal of Environmental Management*, 264, 110446.

Zhao, J.S., Halasz, A., Paquet, L., Beaulieu, C., and Hawari, J. (2002). Biodegradation of RDX and its mononitroso derivative by *Klebsiella pneumoniae* strain SCZ-1. *Applied and Environmental Microbiology*, 68(11), 5336–5341.

Zhao, J.S., Spain, J., and Hawari, J. (2003). Diversity of RDX-transforming bacteria in anaerobic cultures. *FEMS Microbiology Ecology*, 46(2), 189–196.

Zhao, J.S., Spain, J., Thiboutot, S., Ampleman, G., Greer, C., and Hawari, J. (2004). Natural attenuation of explosives by marine sediment bacteria. *FEMS Microbiology Ecology*, 49(3), 349–357.

https://doi.org/10.1142/9789819822249_bmatter

Index

H

I

K

L

M

www.ingramcontent.com/pod-product-compliance
Ingram Content Group UK Ltd.
Pitfield, Milton Keynes, MK11 3LW, UK
UKHW021835100926
8556IPUK00004B/352

9 789819 822232